Springer Theses

Recognizing Outstanding Ph.D. Research

Aims and Scope

The series "Springer Theses" brings together a selection of the very best Ph.D. theses from around the world and across the physical sciences. Nominated and endorsed by two recognized specialists, each published volume has been selected for its scientific excellence and the high impact of its contents for the pertinent field of research. For greater accessibility to non-specialists, the published versions include an extended introduction, as well as a foreword by the student's supervisor explaining the special relevance of the work for the field. As a whole, the series will provide a valuable resource both for newcomers to the research fields described, and for other scientists seeking detailed background information on special questions. Finally, it provides an accredited documentation of the valuable contributions made by today's younger generation of scientists.

Theses are accepted into the series by invited nomination only and must fulfill all of the following criteria

- They must be written in good English.
- The topic should fall within the confines of Chemistry, Physics, Earth Sciences, Engineering and related interdisciplinary fields such as Materials, Nanoscience, Chemical Engineering, Complex Systems and Biophysics.
- The work reported in the thesis must represent a significant scientific advance.
- If the thesis includes previously published material, permission to reproduce this must be gained from the respective copyright holder.
- They must have been examined and passed during the 12 months prior to nomination.
- Each thesis should include a foreword by the supervisor outlining the significance of its content.
- The theses should have a clearly defined structure including an introduction accessible to scientists not expert in that particular field.

More information about this series at http://www.springer.com/series/8790

Yangyang Yang

Artificially Controllable Nanodevices Constructed by DNA Origami Technology

Photofunctionalization and Single-Molecule Analysis

Doctoral Thesis accepted by
Kyoto University, Kyoto, Japan

 Springer

Author
Dr. Yangyang Yang
Department of Chemistry
Graduate School of Science
Kyoto University
Kyoto
Japan

Supervisor
Prof. Hiroshi Sugiyama
Department of Chemistry
Graduate School of Science
Kyoto University
Kyoto
Japan

ISSN 2190-5053
Springer Theses
ISBN 978-4-431-56690-8
DOI 10.1007/978-4-431-55769-2

ISSN 2190-5061 (electronic)

ISBN 978-4-431-55769-2 (eBook)

Parts of this thesis have been published in the following journal articles:

1. "Direct observation of the dual-switching behaviors corresponding to the state transition in a DNA nanoframe" Yang, Y.; Endo, M.; Suzuki, Y.; Sugiyama, H. *Chem. Comm.* **2014**, 50, 4211–4213.
2. "DNA origami technology for biomaterials applications" Endo, M.; Yang, Y.; Sugiyama, H. *Biomater. Sci.* **2013**, 1, 347–360.
3. "Photocontrollable DNA origami nanostructures assembling into predesigned multiorientational patterns" Yang, Y.; Endo, M.; Hidaka, K.; Sugiyama, H. *J. Am. Chem. Soc.* **2012**, 134, 20645–20653.
4. "Single-molecule visualization of the hybridization and dissociation of photoresponsiveoligonucleotides and their reversible switching behavior in a DNA nanostructure" Endo, M.; Yang, Y.; Suzuki, Y.; Hidaka, K.; Sugiyama, H. *Angew.Chem. Int. Ed.* **2012**, 51, 10518–10522.
5. "Programmed placement of gold nanoparticles onto a slit-type DNA origami scaffold" Endo, M.; Yang, Y.; Emura, T.; Hidaka, K.; Sugiyama, H. *Chem. Commun.*, **2011**, 47, 10743–10745.

Supervisor's Foreword

In addition to forming the blueprint for life, DNA molecules can act as building blocks for assembling nanoscale biomaterials. This can be accomplished by exploring their highly specific recognition ability and excellent biocompatibility. DNA nanotechnology has achieved great progress since the emergence of DNA origami technology. Based on this convenient method, various kinds of DNA nanostructures have been developed for multiple applications.

In this thesis, Dr. Yangyang Yang reports how photoisomerization of azobenzene residues in DNA can be used to control the DNA hybridization process. The author has demonstrated the construction of DNA-based molecular switches and the control of the assembly and disassembly of DNA origami. In the first part, which involved construction of DNA nanodevices, a two-step cascading transformation reaction was performed using photoinduced dissociation and G-quadruplex formation. In the second part, the assembly and disassembly of hexagon-shaped DNA origami were regulated reversibly under photoirradiation in a wavelength-dependent manner. For direct visualization of each process, the author used a state-of-the-art high-speed atomic force microscope. The combination of precise manipulation with various kinds of functionalization of different molecules will allow DNA origami to become a powerful tool for studying chemical and biochemical interactions in a defined nanospace.

This thesis clearly demonstrates the potential for the application of DNA nanostructures as nanodevices in various nanosystems.

Kyoto, Japan
August 2015

Prof. Hiroshi Sugiyama

Acknowledgments

Four years of studying and living in Kyoto will be the best memory in my life, including the beautiful views I have seen and every beautiful person I have known.

Foremost, I would like to express my sincere gratitude to my advisor, Prof. Hiroshi Sugiyama, for his accepting me as his doctoral student and the invaluable supervision of my study and research. When I faced with difficulties and failures, his encouragement and gentle guidance always helped me to overcome them finally. When I had problems with life, he helped me go through with them without any doubt.

I also would like to give my great appreciation to Assoc. Prof. Masayuki Endo for the generous and meticulous guidance to my research. I always learned a lot whenever we discussed academic matters.

For my experimental support, I would like express my sincere thanks to Mrs. Kumi Hidaka for AFM imaging with great patience and care. I also want to show my gratitude to Tomoko Emura, who gave me a lot of help in the laboratory.

In addition, I would also like to express my gratitude to the laboratory secretary, Yasuko Niimi, for taking care of me from the first day I arrived in Kyoto. With her assistance and full support, I was able to live in Kyoto with almost no difficulties even when I could not speak any Japanese.

Also I wish to express my thanks to my parents, my husband, and even my baby daughter. With all your full support and complete understanding, I could focus on my research and complete the Ph.D. degree so smoothly.

For financial support, I want to show my great appreciation to the Chinese Scholarship Council (CSC) and the Japan Society for the Promotion of Science (JSPS) for sustaining my life and study at Kyoto University for 4 years.

Contents

Chapter 1
Introduction: Overview of DNA Origami as Biomaterials and Application

1.1 Introduction

Recently, DNA based materials in nanotechnology have been gaining extreme attentions of DNA technologists as well as researchers in other related biosciences. Taking advantages of specific bonding of DNA base pairs and also the biological functions of DNA strands, DNA nanotechnology employ DNA molecules as building blocks to construct programmed nanostructures in various nanometer-scaled sizes and multiple dimensions. The DNA nanotechnology is pioneered by Ned Seeman, who created the geometrical DNA motifs from basic small components to complicated structures, which further promoted the specific concept of structural DNA nanotechnology [1, 2]. The precisely designed and assembled DNA nanostructures can be functionalized with various kinds of molecules and then be applied in molecular mechanics, in synthetic chemistry and also in biological analysis by cooperating with biological methods, and it continues to develop in response to advancing technique demands [3–5].

"DNA origami", a new advent programmed method, allows for folding DNA strands into almost any desired shape and in precisely controlled size [6]. This advanced approach, which is based on well-established DNA nanotechnology, exhibits specially convenient preparing process and has already applied a lot as the nanoscaffolds for material-orientated functionalization and also for single molecule analysis [3–5]. This chapter firstly gives a common introduction of the basics of DNA origami technique such as the design and the programmed construction. Furthermore, some recent representative advances of DNA origami-based nanostructures as biomaterials are highlighted for the applications in the selective functionalization, single molecule analysis, cell-targeted delivery and also the molecular machine.

© Springer Japan 2015
Y. Yang, *Artificially Controllable Nanodevices Constructed by DNA Origami Technology*, Springer Theses, DOI 10.1007/978-4-431-55769-2_1

1.2 2D DNA Origami Nanostructures

DNA origami, a method to create two dimensional (2D) DNA nanostructures in ∼ 100 nm size was developed by Rothemund in 2006 [6]. A long single stranded DNA (M13mp18, 7249 bases) is mixed with a series of sequence-complementary strands called staple strands (∼ 200 strands, each ∼ 32-mer) and then annealed from 95 °C to room temperature over 2 h. The predesigned nanostructure will be well formed, as illustrates in Fig. 1.1a. Besides the rectangular-shaped structure, triangles, smiley face and five-pointer star were successfully constructed. The assembled structures can be easily confirmed by atomic force microscopy (AFM), in which even the hybridized DNA duplexes can be clearly observed in high resolution (Fig. 1.1c).

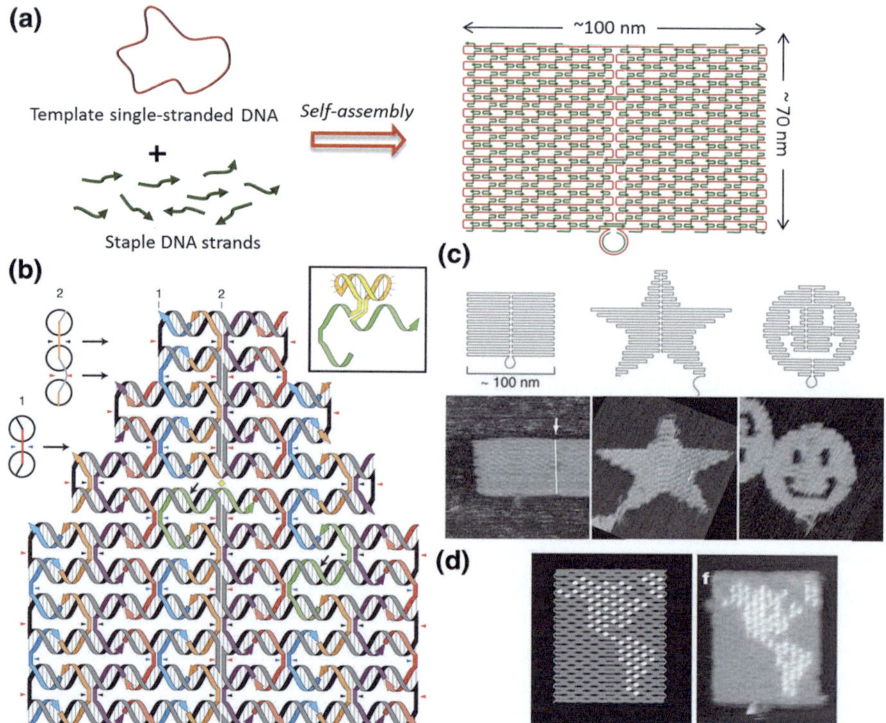

Fig. 1.1 Introduction of DNA origami method. **a** A long single-stranded DNA employed as template assembled with hundreds of staple DNA strands to form a rectangular-shaped nanostructure in a size of ∼ 100 nm × ∼ 70 nm. **b** Detailed design of a DNA origami: the short staple strands connect the adjacent duplexes with crossovers. *Inset* a DNA hairpin structure as a topological marker. **c** Design and AFM images of 2D DNA origami structure in various shapes. **d** Drawing of a hemisphere pattern on a DNA origami with hairpin DNAs (*white dots*) and its corresponding AFM image. Reprinted by permission from Macmillan Publishers Ltd: [Nature] (6), copyright (2006)

The basic design rules are demonstrated in Fig. 1.2b. The adjacent double-stranded DNAs (dsDNAs) are connected by staple strands as crossovers. In this design, the geometry of the double helices involved has three helical rotations for 32 base pairs. Two neighboring crossovers of the central dsDNA in an arrangement of three adjacent dsDNAs should be located at the opposite sites (180° rotated, 0.5 turns). Therefore, the crossovers should be separated by 16 base pairs (1.5 turns). To obtain a planar 2D structure, the multiple staple strands pulling the long strand should be arranged following the same rule. Hundreds of staple strands also afford the possibilities of locating hairpin DNAs as topological markers at desired positions, which are imaged as a dot on AFM images. In principle, the hairpin DNA is placed at a position eight base pairs from the crossover (270° rotation) where the hairpin DNA is located perpendicular to the surface of the origami structure. The distance between neighboring staples strands was calculated around 6 nm, representing that the neighboring hairpin DNAs can be imaged as distinct dots by AFM imaging. As a result, the hairpin DNAs can be employed to constitute various patterns directly on DNA origami surface, such as the map of a hemisphere as shown in Fig. 1.1d. Besides the introduction of the hairpin DNAs into the origami structures, functional molecules and nanoparticles can also integrated into the nanostructures by the modification with staple strands at predetermined positions or by the specific interactions with the specially designed staple strands.

A remarkable property of DNA origami is that each position on the structures possesses one piece of specific sequence information which can be taken as an address no matter the shape variations. Before the advent of DNA origami method, it is quite a challenge to uniformly create nanostructures in a defined size ~ 100 nm through the assembly of small DNA components. Right now, a big progress is accomplished, which enable no matter the shape design or the size limitation. By using this method, the size of the various geometric DNA origami structures are determined by the template strand: the long single stranded DNA. Various single-stranded DNAs were isolated to serve as the fresh template for the assembly of different nanostructures [7, 8]. A strategy of using DNA tiles (17 × 16 nm) instead of staples has also been developed, which allows size expansion by the introduction of 25–56 DNA tiles [9].

1.3   Programmed Arrangements of Large Sized DNA Nanostructures

The size of single DNA origami structure is constrained around 100 nm because of the length limitations of the single stranded DNA template. To obtain larger-sized structures based on origami method, other techniques are required, which can integrate more complicated functions together.

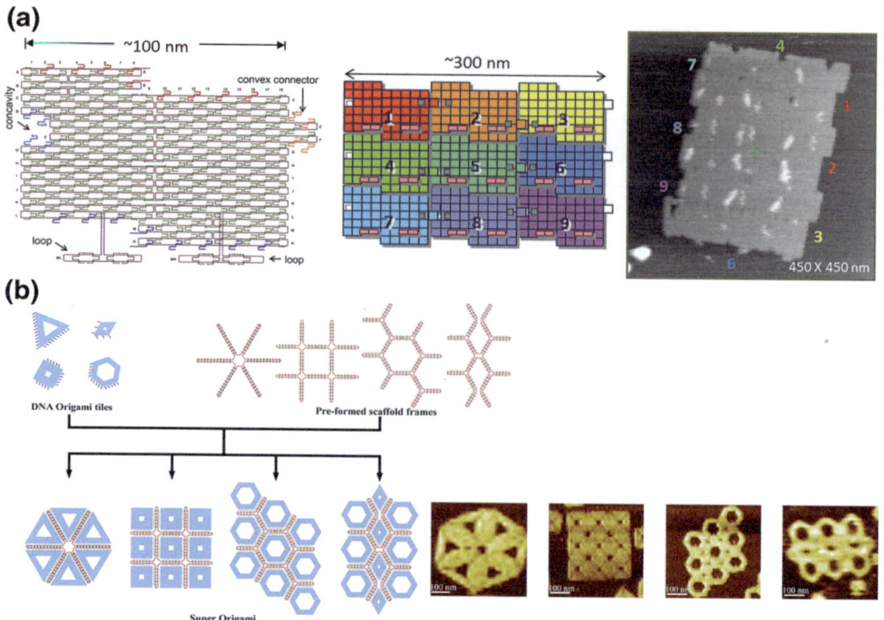

Fig. 1.2 Programmed self-assembly of DNA origami. **a** A DNA origami structure with a concavity and a convex as a "DNA Jigsaw piece" for 2D assembly. A 3 × 3 assembly of nine predesigned origami tiles and its AFM image. Reprinted with permission from (Rajendran, A.; Endo, M.; Katsuda, Y.; Hidaka, K.; Sugiyama, H. *ACS nano* **2010**, *5*, 665–671) Copyright (2010) American Chemical Society. **b** Programmed assembly of "Superorigami" structures by using multiple DNA origami structures scaled up by scaffold frames and their corresponding AFM images. Reprinted with permission from (Nano Letters). Copyright (2011) American Chemical Society

Our group have explored techniques called "Jigsaw pieces" to assemble 2D DNA tiles into larger 2D programmed horizontal or vertical patterns [10, 11]. The assembling tile was designed in a rectangular shape (Fig. 1.2a), in which the tiles can assemble together along the helix axis direction via at the edges. Moreover, specific concave and convex connectors were introduced into the tile to align multiple rectangles precisely with each other. Multiple rectangular tiles can assemble together by both shape and complementary strands which were extended from the staple strands located at the lateral edges of tiles. Nine different tiles were designed and prepared separately. Firstly, three tiles were programmed to assemble together in either vertical or horizontal directions. Secondly, three sets of trimers were finally assembled into a 3 × 3 large structure (300 nm × 240 nm, AFM image in Fig. 1.2a), which is nine-fold larger than single DNA tile.

Rothemund and his co-workers developed a programmed assembly system by regulating the positions of adhesive π-stacking ends for selectively connect rectangular tiles [12]. Besides, Yan's group introduced a template-assisted assembly of different shaped DNA origami monomers into large structures in various patterns

(Fig. 1.2b). The assembling unit assembled with the templates prepared with programmed patterns in a shape complementary rule. As shown in Fig. 1.2b, triangles, rectangles and hexagons were scaled up together in a predesigned manner [13].

The programmed assembly of larger sized nanostructures can be realized by using the above strategies, which can scale up assembling monomers into predesigned patterns and defined size. The extended DNA nanostructures exhibited more address information and provide more opening positions for further functionalization.

Besides using the long single-stranded DNA as scaffold for programmed assembling which requires the helps of hundreds of short "staple" strands, another 'single-stranded tile' (SST) approach [14] was developed by Yin and coworkers. As shown in Fig. 1.3a, a SST motif consists of 4 domains in a total length of 42 nucleotides, which composed of concatenated sticky ends assembled with four

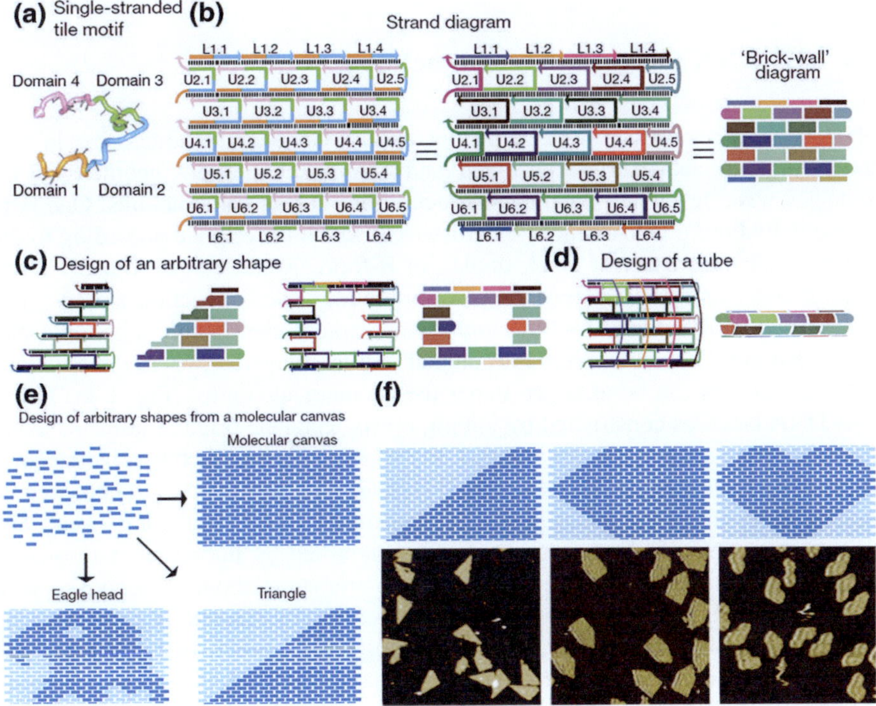

Fig. 1.3 **a** Single-stranded tile motif consisting of four domains in 2 parallel pairs as referred from Ref. [16]. **b** Schematic diagram of rectangular 'brick wall' in 6 helix × 8 helical turn (6H × 8T). Each strand has a unique sequence. **c** An arbitrary shape, triangle (*left*) or rectangular ring (*right*), is formed by selecting proper subset of SST tiles from the groups in (**b**). **d** A tube structure with prescribed width and length. **e** Taken each SST tile as a molecular pixel, molecular canvas or derived eagle head and triangle shaped structured can be designed freely. **f** Simple shapes were designed and constructed using a molecular canvas. Reprinted by permission from Macmillan Publishers Ltd: [Science] (15), copyright (2012)

neighboring strands [15]. A 'brick wall' pattern is formed by arranging a series of distinct SST tiles into lattices (Fig. 1.3b), in which one SST tile can be folded into a 3 nm-by-7 nm tile and associate with four neighboring tiles acting as a pixel. Based on this strategy, any arbitrary shapes of nanostructures using different combinations of SST tiles can be assembled even a tube by connecting half-tiles on the top and bottom boundaries into full tiles (Fig. 1.3c, d). Moreover, a predesigned rectangular lattice can also be viewed as a 'molecular canvas' (shown in Fig. 1.3e) in which each SST tile serves as one 'molecular pixel'. By selecting constituent 'molecular pixels' on the canvas, such as eagle head and triangle (Fig. 1.3e), multiple shapes are formed and constructed, as illustrated in Fig. 1.3f. This SST strategy allows the self-assembly of desired complex structures without the specific scaffold routing design. No matter SST method or conventional DNA origami strategy, both shows that a vast space for developing various approaches for the construction of nucleic acid nanostructures.

1.4 3D DNA Origami Nanostructures

DNA origami nanostructures can be not only extended into larger size in two dimension, but also can be designed and assembled into 3D geometry. Two strategies were developed for the construction of 3D origami structures. One is to bundle neighboring DNA helices by crossovers which is designed according to the structural characteristics of DNA duplex in B-form. Another is to join 2D DNA origami domains into 3D layouts by corresponding interconnection strands. The former method developed by Shih and his coworkers, the relative adjacent DNA helices are joined by crossovers into tubular and multilayered, where the positions of the crossovers can be arranged under the geometrical control (Fig. 1.4a) [16].

A DNA box was constructed by joining up six rectangle origami domains using interconnecting strands (shown in Fig. 1.4b), which was confirmed by cryo-electron microscopy (cryo-EM) as well as AFM [17]. Interestingly, the lid of 3D box can be initiated to open by the addition of complementary strand employing strand displacement strategy. The opening event was monitored by fluorescence resonance energy transfer (FRET) (Fig. 1.4c). Based on similar method, DNA boxes with controlled inside and outside interface were successfully created by regulating the directions of crossovers at the connection edges [18]. A tetrahedral structures was constructed by connecting four aligned triangles, which were pretreated with an M13mp18 scaffold strand without the formation of independent 2D pieces [19].

Our group introduced a series of hollow prism structures containing different numbers of 2D origami pieces [20]. Interestingly, the high-speed AFM imaging captured the opening events of the hollow 3D tube-like structure in second time-scale. Recently, based on SST strategy, Yin's group developed 3D complex structures using short synthetic DNA strands called "DNA bricks" [21]. Hundreds of unique bricks self-assembled into prescribed 3D shapes just like Legos in one-step.

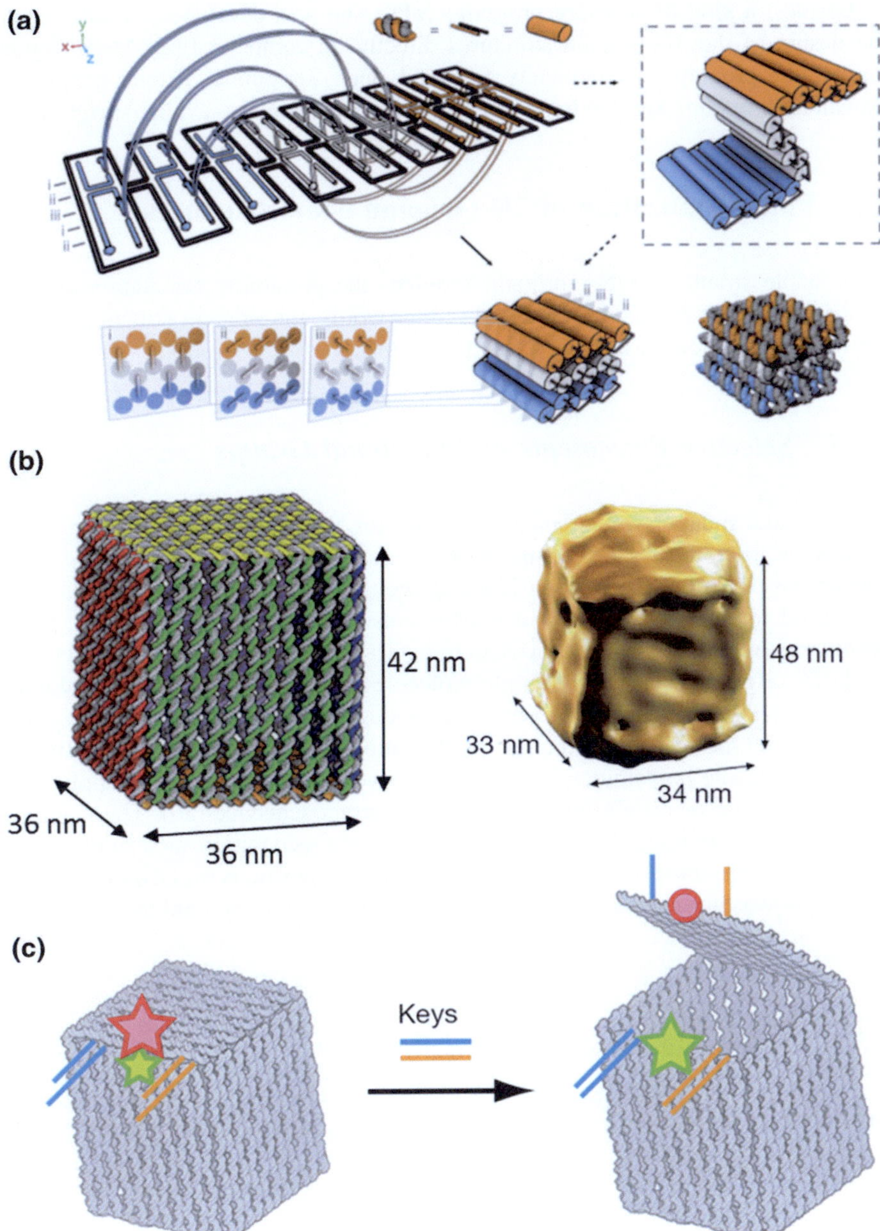

Fig. 1.4 3D DNA box with controlled opening by using strand displacement strategy. **a** 2D pleated structure folded into a 3D multilayered structure using staple strands connecting adjacent layers. Reprinted by permission from Macmillan Publishers Ltd: [Nature] (17), copyright (2009). **b** A 3D DNA box was formed and confirmed by cryo-EM. **c** Controlled opening of the box lid using a selected DNA strand as a key. Lid opening event was monitored by FRET. Reprinted by permission from Macmillan Publishers Ltd: [Nature] (18), copyright (2009)

Besides, a kind of software program: caDNAno was developed, which enable the design of the 3D origami structures directly on computer [22]. More importantly, this program is completely free and open for public. People can freely download it online. And the basic tutorials for starters are also available right now.

1.5   Functionalization of 2D Origami Nanostructures

Each staple strand on DNA origami provides the possibility for addressing and modification with functional groups. In other words, the origami structure affords the desired functionalization at defined positions.

1.5.1   Selective Placements of Functional Groups

Ligands and aptamers are popular particles for the conjugation between DNA strands and proteins, indicating that proteins can selectively attached on the DNA origami by specific recognitions [23–26]. For instance, SNAP-tag and halo-tag were used for the placement of fusion proteins on the DNA origami [27]. Yan's group employed single-stranded DNAs to detect target RNA molecules directly on the DNA origami surface at the single-molecule level by AFM [28]. By employing the streptavidin-biotin labeling, our group designed a five-well nanostructure, which is used for the characterization of alkylation of the sequence selective ligand: pyrrole-imidazole polyamide [29]. The same five-well frame was also applied for the investigation of the Zn-finger protein with target sequence [30]. These studies sufficiently demonstrate that the designed sequence recognition molecules can be used for the placement of proteins on predetermined positions on DNA origami. It is also shown that the AFM imaging can be used to directly confirm functionalization in single molecule level.

1.5.2   Single-Molecule Chemical Reaction

Not only the biomolecules, small organic molecules for specific chemical reactions can be performed and imaged at the local positions of DNA origami by using AFM at single molecule level. Chemical reactions: reductive cleavage of disulfide bonds and oxidative cleavage of olefin by single oxygen were carried out on the DNA origami surface [31]. Besides, amide bond formation and click reactions were also performed on the DNA origami scaffold. Reactive groups: azido, amino and alkyne groups conjugated with staple strands were located at different positions of DNA origami. Three coupling reactions were successively carried out using the

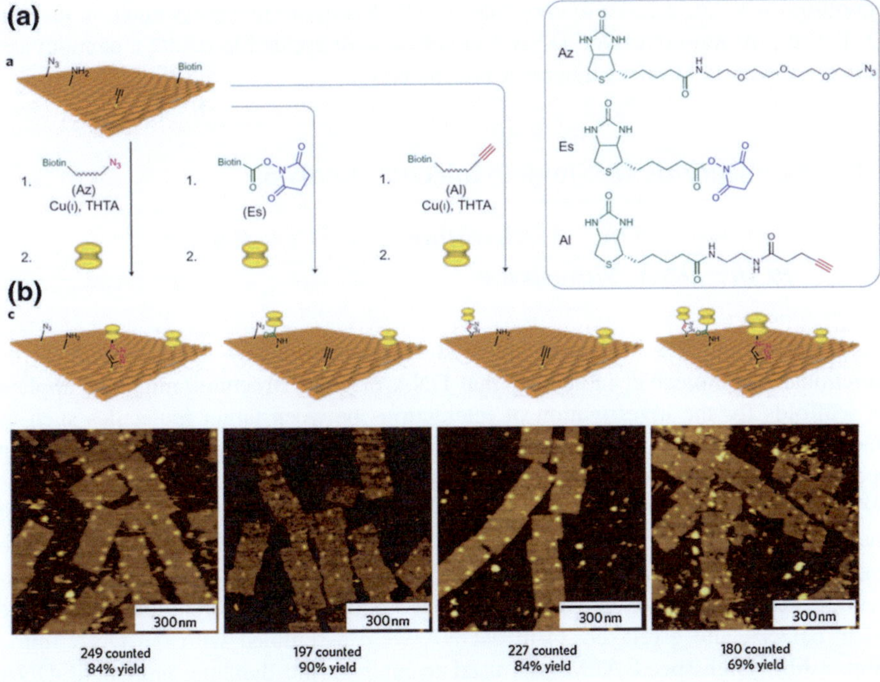

Fig. 1.5 Single chemical reaction on DNA origami. **a** Three active groups: azido, amino and alkyne groups were selected and incorporated together into one single DNA tile by modification with staple strands. The coupling reactions of functional groups attached by biotin were carried out on DNA tile, which were visualized by the specific binding of streptavidin-biotin. **b** AFM images of three reactions performed at different positions. Reprinted by permission from Macmillan Publishers Ltd: [Nature Nanotechnology] (32), copyright (2010)

biotin-attached functional groups (Fig. 1.5). Both of the bond cleavage and the bond formation were characterized by employing streptavidin binding and confirmed by AFM imaging. These chemical reactions all proceeded quantitatively on the defined positions of the DNA origami in high yield.

1.5.3 Selective Modifications with Nanomaterial

Other kinds of nanomaterials can also be integrated into the DNA nanostructures. Gold nanoparticles (AuNPs) coupled with staple strands were selectively placed on DNA origami [32, 33]. The AuNPs can also be encapsulated into a DNA origami cage with addressable and spatial control [34]. Alternatively, thiol-group tethered staple strands can be used to accumulate the AuNPs into DNA origami at pre-designed positions in various patterns [35, 36]. Two DNA-modified carbon nanotubes were arranged into both sides of DNA origami in a cross junction fashion [37]. The position-controlled DNA nanotubes were applied to create a nanodevice, which showed field-effect transistor-like behavior.

1.6   Applications to Single Molecule Analysis

1.6.1   *Control of DNA Methylation and DNA Repair in the DNA Nanospace*

Nanometer-sized DNA origami can be clearly visualized by AFM as well as the functionalized molecule, indicating that DNA origami structures might be applied as scaffolds for the investigation of interactions between target molecules such as the DNA-enzymatic reactions. DNA-targeted enzymes often require bending specific DNA strands to facilitate the reaction. The dsDNA will be bended by 55–59° during the methyl-transfer reaction by DNA methylation enzyme *Eco*RI methyltransferase (M.*Eco*RI) [38]. To examine the DNA structural effect on the methylation reaction, our group designed a 2D DNA frame containing four connection sites at the inner side (Fig. 1.6a). Two dsDNAs in different lengths: a tense 64-m dsDNA and a relaxed 74-m dsDNA were assembled with the DNA frame (Fig. 1.6b). High-speed AFM was used to analyze the dynamic enzymatic-DNA

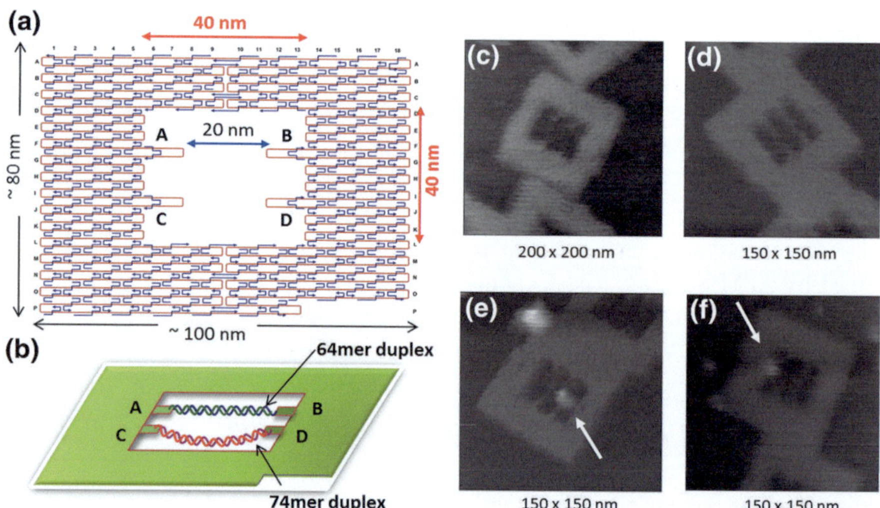

Fig. 1.6 Control of DNA methylation reaction on a DNA frame. **a** Design of the DNA frame containing four connection sites in the inner cavity. **b** Two dsDNAs in different length: AB = tensed 64-m dsDNA and CD = relaxed 74-m dsDNA carrying recognition sequence for M. *EcoRI*, were incorporated into the frame. AFM image of empty DNA frame (**c**), two dsDNA cooperated DNA frame (**d**), and M. *EcoRI* bound to dsDNA of AB (**e**) and CD (**f**), respectively. Reprinted with permission from (Endo, M.; Katsuda, Y.; Hidaka, K.; Sugiyama, H. *J. Am. Chem. Soc.* **2010**, *132*, 1592-1597.). Copyright (2010) American Chemical Society

interaction directly on mica surface. The obtained AFM images revealed that methylation reaction of M. *EcoRI* preferred to occur in the relaxed 74-m dsDNA rather than in the 64-m dsDNA [39]. The results confirmed that the structural flexibility of bending dsDNA is important for the methyl-transfer reaction with M. *EcoRI*.

The same DNA frame was also applied for the analysis of DNA repair reactions and directly visualized by high-speed AFM. DNA base excision repair enzymes: 8-oxoguanine glycosylase [40] and T4 pyrimidinebdimer glycosylase [41] were incorporated into the DNA frame containing various dsDNAs with a damaged base [42]. The DNA repair reactions were examined on the DNA frame in single molecule level. The relaxed 74-m dsDNA trapped the DNA with the cleavage by $NaBH_4$ and exhibited higher cleavage efficiency than 64-m dsDNA. Then the dynamic movements of the enzymes on dsDNA in DNA frame were also directly observed by using high-speed AFM.

DNA origami exhibited as excellent scaffold to integrate with functional molecules for single molecule analysis and the dynamic interactions of active molecules can be directly visualized in the designed nanoscale space by using high-speed AFM imaging.

1.6.2 Visualization of DNA Structural Changes

Not only the DNA-enzymatic interactions can be visualized, but also the structural interaction between DNA strands can be directly observed on DNA origami structures. The reversible G-quadruplex formation and disruption was manipulated on a single DNA frame by using potassium as stimuli [43].

Based on the same frame structure, two G-rich fragments were introduced into the central positions of two dsDNA (both 64-m) separately: three-G-tracts were located in the upper dsDNA whereas the lower dsDNA contained a single G-tract [44] (Fig. 1.7). After the addition of KCl, the two dsDNA were associated together in an X-shape because of the formation of G-quadruplex structure. AFM imaging was employed for the confirmation of the structural changes between two dsDNAs and the yield of the association efficiency was evaluated manually. Furthermore, the real-time dynamic structural changes indicating the formation and disruption of interstrand G-quadruplex were directly observed in different observation buffers with/without K^+ by high-speed AFM. This was the first report of the real-time direct observation of DNA structural changes in limited nano-space. This method can be applied for the analysis of various kinds of nucleic acid-related structural changes in single molecule level.

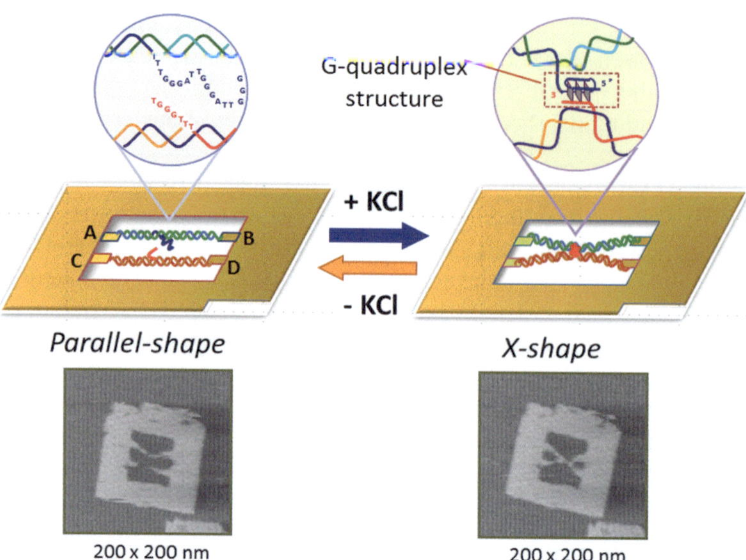

Fig. 1.7 Visualization of the structural changes of two dsDNAs corresponding to the reversible G-quadruplex formation and disruption in the DNA frame. In the presence of KCl, the two dsDNAs were separated. The two dsDNAs formed X-shape at the central positions via G-quadruplex formation by addition of K^+. Reprinted with permission from (Sannohe, Y.; Endo, M.; Katsuda, Y.; Hidaka, K.; Sugiyama, H. *J. Am. Chem. Soc.* **2010**, *132*, 16311–16313). Copyright (2010) American Chemical Society

1.6.3   Single Molecule Fluorescent Imaging

Besides AFM and transmission electron microscopy (TEM), single-molecule fluorescence microscopy also plays an important role particularly for the investigation of complex DNA nanostructures [45]. Taking advantage of the precise addressability of DNA origami each staple strand can be distinguished as a unique attachment point for various nanoobjects such as different fluorophores. Tinnefeld and his coworkers presented a nanoscopic molecular ruler based on DNA origami methods as a general platform for super-resolution microscopy using different imaging techniques [46]. As shown in Fig. 1.8a, a rectangular DNA origami nanostructure containing two fluorescently labeled staple strands at two specific positions (F in black circle), where the distance between them were below the diffraction limit. The assembled structures were immobilized and easily confirmed by AFM. Total internal reflection fluorescence (TIRF) microscopy imaging resulted in a diffraction-limited image in which the emission patterns of two fluorophores were overlapped. However, single fluorophore were identified by using super-resolution fluorescence microscopy (Fig. 1.8a, b). In Fig. 1.8b, the blink microscopy was employed. The fluorescent labeled DNA origami served as rulers

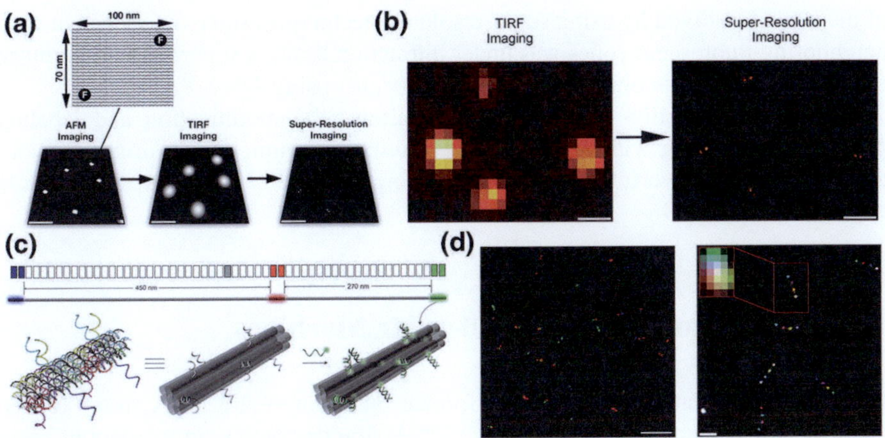

Fig. 1.8 Schematic drawings of DNA molecular nanoscopic ruler. **a** A rectangular DNA origami structures labeled with two fluorophores at two positions below diffraction limit. AFM was used for the confirmation of the nanostructures. TIRF imaging and super-resolution imaging were employed to record the emission patterns of two fluorophores. **b** *Left* TIRF image of surface-immobilized DNA origami labeled with two ATTO655 fluorophores resulting overlapping point-spread functions; *right* super-resolution image of same region using blink microscopy. [DNA Origami as a Nanoscopic Ruler for Super-Resolution Microscopy/Steinhauer, C.; Jungmann, R.; Sobey, T. L.; Simmel, F. C.; Tinnefeld, P./Angew Chemie International Edition, 48/47. Copyright (c) [2009] [copyright owner as specified in the Journal]. **c** Schematics of barcode with a segment diagram on the *top* and a 3D view at the *bottom* and the specific labeled zone in a 3D illustration. **d** *Left* a representative superimposed TIRF image of barcode containing 27 equimolar mixtures of barcode species; *right* a super-resolution fluorescent image of DNA-barcodes. Reprinted by permission from Macmillan Publishers Ltd: [Nature Chemistry] (48), copyright (2012)

for the calibration of super-resolution microscopy and also might be used as quantification standards for other spectroscopic techniques.

A series of geometrically fluorescent barcodes constructed from DNA nanostructures which were assembled with fluorophore-modified DNA strands were developed by Yin and coworkers [47]. A six-helix bundle DNA nanorod was designed and assembled (Fig. 1.8c) where three zones were chosen for the hybridization with three different fluorescent-labeled DNAs, which result the combinations of $3^3 = 27$ barcodes in spatially distribution pattern. After the assembling of DNA nanorods with the specified fluorophores, the barcode system was confirmed by TIRF. A representative image of a pool of all 27 members of barcode family was shown in Fig. 1.8d (left), indicating that the long DNA nanorods exhibited a reliable system to construct geometrically encoded barcodes using fluorescent-imaging technique. By increasing the number of fluorescent zones, the complexity of barcodes can also be enhanced. Meanwhile the resolution of them can be solved by using super-resolution techniques since the space between neighboring fluorescent zoncs was under diffraction limit. A super-resolution image of a five-barcode was obtained in Fig. 1.8d (right) using DNA-PAINT [48].

The programmability of DNA origami allows the modification and labeling precisely. Integrating with high spatial resolution technique, DNA origami gradually becomes a powerful method for the single molecule analysis in nanoscale applications.

1.7 Applications to DNA Molecular Machines

DNA based molecular machine is performed by adding extra DNA strand (called "fueling strand") and replacing specific DNA strands containing a "toehold" segment selectively to initiate the mechanical movements in the nanosystem. During the strand displacement, the thermodynamic stabilization energy works for the hybridization of new duplex and control the mechanical movements of DNA machines. By employing this strategy, a DNA based molecular tweezer was created by Yurke and co-workers to perform close-open motions [49]. Another kind of DNA walking device can move autonomously by the cleavage of DNA strand by DNA nicking enzyme [50].

Seeman and coworker developed a rotational mechanical device called "PX-JX2 device", in which the two adjacent ends of dsDNAs can be regulate to rotate in either the PX state and JX2 state by hybridization topology [51–53]. Later two PX-JX2 devices were introduced on the DNA origami scaffold and dynamic patterning were captured on the DNA origami [54]. And a DNA walker was developed to operate the assembly line on DNA origami, in which three PX-JX2 device were also incorporated into the DNA origami framework and then delivered AuNPs to walker which picked up the AuNPs at specific positions [55]. A DNA nanomachine called "DNA spider" were reported by Yan and coworkers [56], in which the programmed operations such as "start", "walk" and "stop" can be incorporated into predesigned track.

A DNA transportation device based on DNA origami was constructed which allow the walking movements along the linear track [57]. A linear track (multiple single stranded DNAs, stators) was introduced to the surface of a 2D DNA tile to guide the stepwise movement of the walking strand (Fig. 1.9a). The cleavage of walker-stator duplex at specific sequence position by nicking enzyme Nt.BbvCI [58] removed a part of stator strands. The walker strand would search for neighboring intact strand and hybridize with it fully by branch migration (Fig. 1.9b). The time-dependent movement of the walker was analyzed by AFM, which was imaged as a single spot representing the walker-stator duplex on the DNA tile. Furthermore, the stepwise movement of the walker was directly observed by high-speed AFM

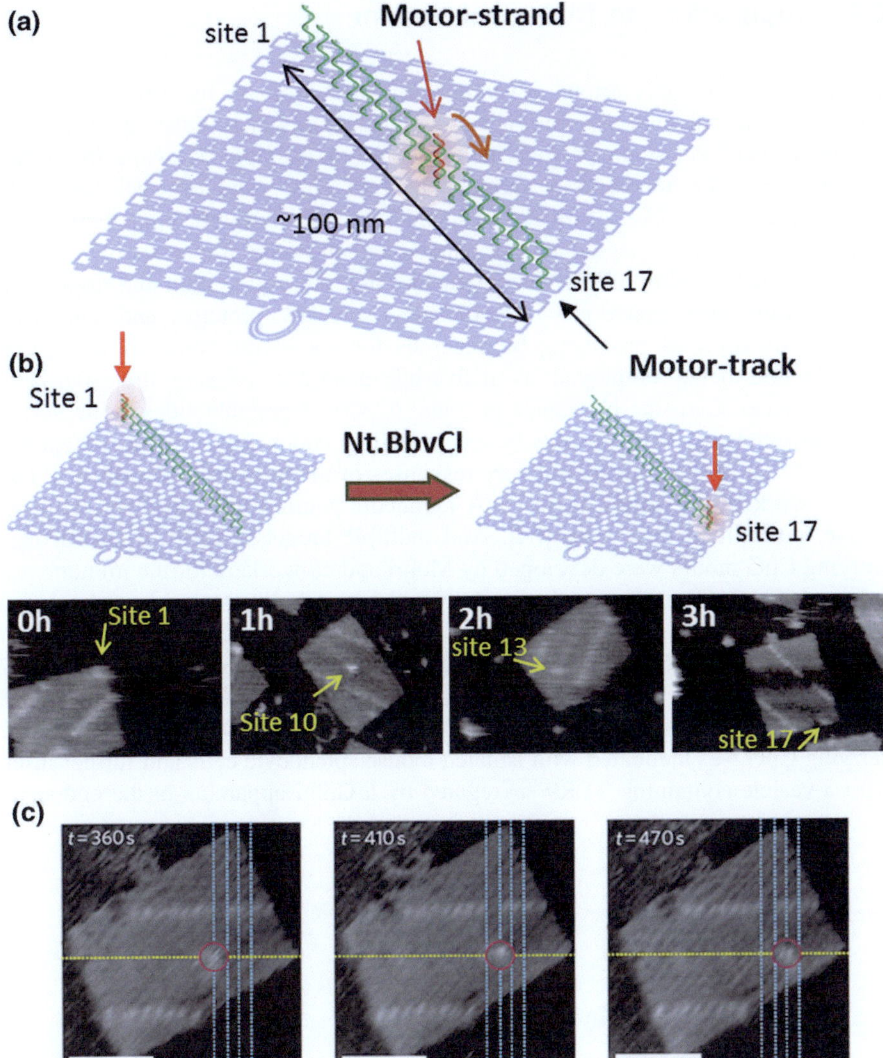

Fig. 1.9 DNA motor on the DNA origami. **a** A 2D DNA tile containing a linear motor track to guide the movement of the motor-strand. **b** Time-dependent movement of a DNA motor by Nt. BbvCI. **c** Step-wise movement of DNA motor observed by high-speed AFM. Scale bar: 50 nm. Reprinted by permission from Macmillan Publishers Ltd: [Nature Nanotechnology] (58), copyright (2011)

(Fig. 1.9c). Our group later designed a complicated walking system for controlling the walking directions using a branched track and controllable blocking strands. The walker's movement along the branched track to defined destination was also directly visualized by high-speed AFM [59]. This method can also be applied to the cargo transportation or perform specific task in limited nanospace.

1.8 Applications to Biological System

The above mentioned various advantages of DNA origami nanostructures have showed great potential for biological applications, which have already been extended to cellular studies. There have been a few examples showing that DNA nanostructures are resistant to various kinds of endo- and exonucleases [60]. Meldrum and co-workers further illustrated that DNA origami constructs could keep integrity without degradation or damage in cell lysate of a series of cell lines [61]. The high stability of DNA nanostructure in biological system and favourable compatibility with functional biomolecules such as proteins and aptamers, demonstrate itself as promising biomaterials for the investigation of living cell analysis and being employed as delivering platforms of safe drug delivery. Unmethylated Cytosine-phosphate-guanine (CpG) oligonucleotides with strong immunostimulatory activities can be recognized by endosomal Toll-like receptor 9 (TLR9) to induce immunostimulatory responses of the immune cells [62, 63]. Fan and co-workers developed a 3D DNA tetrahedra bearing CpG motifs for noninvasive intracellular delivering [64]. And multiple branched DNA nanostructures carrying CpG motifs were developed by Mohri and co-workers for the immune cell deliver [65]. CpG motifs can be accumulated into a hollow tube-shaped origami (Fig. 1.10), which were introduced into the mammalian cells for the investigation of immune responses [66]. The hollow tube containing as many as 62 inner and 62 outter binding sites (Hs) for CpG and anchor sequence was designed in a size of ∼80 nm × 20 nm, which is fit with the cellular uptake size. The hollow tube carrying CpG was incubated with isolated mouse splenocyte cells and further fused with a vesicle containing TLR9 segregated by a Golgi apparatus in the endocytic

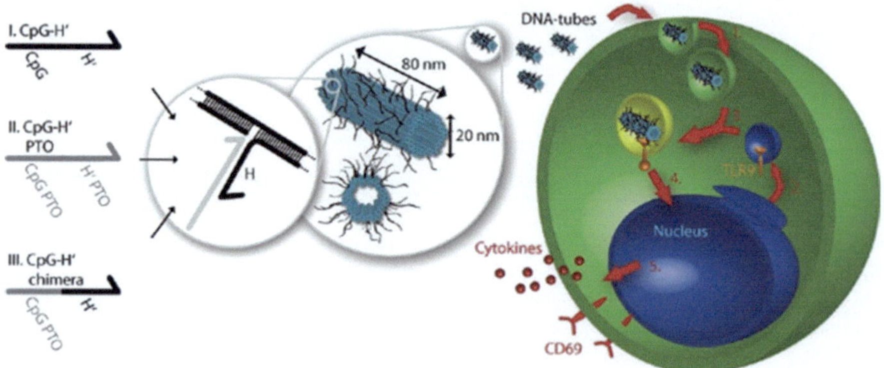

Fig. 1.10 Schematic design of tube-shaped DNA origami and predicted endocytotic pathway. *Left* Three different types of CpG-H'designed to hybridize with CpG. *Right* endocytic pathway of tube origami holding CpG with immune cells and subsequently stimulating the immune responses. Reprinted with permission from (Li, J.; Pei, H.; Zhu, B.; Liang, L.; Wei, M.; He, Y.; Chen, N.; Li, D.; Huang, Q.; Fan, C. *ACS nano* **2011**, *5*, 8783–8789). Copyright (2011) American Chemical Society

pathway. The immune signalling cascade in cell was induced by the recognition of CpG sequences on hollow tube by TLR9. Cytokines and CD69 were further expressed to stimulate the immune response.

Douglas and co-workers designed a 3D hexagonal barrel encapsulated with biomolecules or nanoparticles [67]. The hexagonal barrel was firstly shut by the special DNA latches, which can be unlocked to open by the specific cell-surface proteins (antigen) and the payloads inside will be delivered. The hollow inside can be loaded with different kinds of payloads such as active proteins and nanoparticles. The aptamer locking mechanism can be designed to work in a logic-gated manner according to the different combinations of molecular inputs from target cells. Furthermore, this technique can be used to interface with cells and stimulate the cell signaling responses in an inhibition or activation manner. This programmed DNA-origami based delivery system provides the possibilities of future medical applications such as targeted therapies.

1.9   Conclusion and Prospects

As a bottom-up strategy, DNA origami provides the researchers in nanotechnology a convenient method to realize the construction of nanometer-sized nanostructures in multi-dimensional based on structural DNA nanotechnology. This method affords the possibilities of design and constructs arbitrary nanostructures to satisfy with different purposes. And also this method greatly reduces the experimental labor, eliminates uncertainties to obtain structures in relatively high yield, which greatly increases the efficiency for applications such as single molecule imaging, dynamic mechanical analysis, and cell-targeted delivery and also have already been served for top-down nanotechnology. The constructed nanostructures with pre-designed geometries and specific functionalization exhibit various advantages when compared with other kinds of known biomaterials. As the size of DNA origami is compatible with cellular uptake, the cell-targeted delivery of DNA origami in various shapes with selected functionalization is expected to be applicable. DNA origami can also be employed as assembling components for higher-leveled functionalities with programmed arrangements. The organization and placement of small molecules into confined nanospace is still a challenge. If the preparation of functionalized DNA origami structures can be modified and purified to exclude incomplete portions, the applications of DNA origami method will be greatly improved and become more popular.

DNA origami allows for the addressing functional molecules precisely at pre-determined positions. The nanometer-sized nanostructures are able to accumulate specific synthetic molecules for chemical and biological reactions in nanospace. It is also can be employed as nanoscaffold for the analysis of dynamic molecular behaviors under nanometer resolution in single molecule level. This technology also opens the way to express the complex functionality of the programmed organization of different modules seen in living systems.

References

1. Seeman NC (1982) J Theor Biol 99:237–247
2. Seeman NC (2003) Nature 421:427–431
3. Endo M, Sugiyama H (2009) Chem-Eur J Chem Biol 10:2420–2443
4. Rajendran A, Endo M, Sugiyama H (2012) Angew Chem Int Ed 51:874–890
5. Torring T, Voigt NV, Nangreave J, Yan H, Gothelf KV (2011) Chem Soc Rev 40:5636–5646
6. Rothemund PWK (2006) Nature 440:297–302
7. Högberg B, Liedl T, Shih WM (2009) J Am Chem Soc 131:9154–9155
8. Pound E, Ashton JR, Becerril HcA, Woolley AT (2009) Nano Lett 9:4302–4305
9. Zhao Z, Yan H, Liu Y (2010) Angew Chem Int Ed 49:1414–1417
10. Endo M, Sugita T, Katsuda Y, Hidaka K (2010) Chem–Eur J 16:5362–5368
11. Rajendran A, Endo M, Katsuda Y, Hidaka K, Sugiyama H (2010) ACS Nano 5:665–671
12. Woo S, Rothemund PWK (2011) Nat Chem 3:620–627
13. Zhao Z, Liu Y, Yan H (2011) Nano Lett 11:2997–3002
14. Wei B, Dai M, Yin P (2012) Nature 485:623–626
15. Yin P, Hariadi RF, Sahu S, Choi HMT, Park SH, LaBean TH, Reif JH (2008) Science 321:824–826
16. Douglas SM, Dietz H, Liedl T, Hogberg B, Graf F, Shih WM (2009) Nature 459:414–418
17. Andersen ES, Dong M, Nielsen MM, Jahn K, Subramani R, Mamdouh W, Golas MM, Sander B, Stark H, Oliveira CLP, Pedersen JS, Birkedal V, Besenbacher F, Gothelf KV, Kjems J (2009) Nature 459:73–76
18. Kuzuya A, Komiyama M (2009) Chem Commun 28:4182–4184
19. Ke Y, Sharma J, Liu M, Jahn K, Liu Y, Yan H (2009) Nano Lett 9:2445–2447
20. Endo M, Hidaka K, Kato T, Namba K, Sugiyama H (2009) J Am Chem Soc 131:15570–15571
21. Ke Y, Ong LL, Shih WM, Yin P (2012) Science 338:1177–1183
22. Douglas SM, Marblestone AH, Teerapittayanon S, Vazquez A, Church GM, Shih WM (2009) Nucleic Acids Res 37:5001–5006
23. Chhabra R, Sharma J, Ke Y, Liu Y, Rinker S, Lindsay S, Yan H (2007) J Am Chem Soc 129:10304–10305
24. Shen W, Zhong H, Neff D, Norton ML (2009) J Am Chem Soc 131:6660–6661
25. Rinker S, Ke Y, Liu Y, Chhabra R, Yan H (2008) Nat Nanotechnol 3:418–422
26. Kuzuya A, Kimura M, Numajiri K, Koshi N, Ohnishi T, Okada F, Komiyama M (2009) ChemBioChem 10:1811–1815
27. Saccà B, Meyer R, Erkelenz M, Kiko K, Arndt A, Schroeder H, Rabe KS, Niemeyer CM (2010) Angew Chem Int Ed 49:9378–9383
28. Ke Y, Lindsay S, Chang Y, Liu Y, Yan H (2008) Science 319:180–183
29. Yoshidome T, Endo M, Kashiwazaki G, Hidaka K, Bando T, Sugiyama H (2012) J Am Chem Soc 134:4654–4660
30. Nakata E, Liew FF, Uwatoko C, Kiyonaka S, Mori Y, Katsuda Y, Endo M, Sugiyama H, Morii T (2012) Angew Chem Int Ed 51:2421–2424
31. Voigt NV, Torring T, Rotaru A, Jacobsen MF, Ravnsbaek JB, Subramani R, Mamdouh W, Kjems J, Mokhir A, Besenbacher F, Gothelf KV (2010) Nat Nanotechnol 5:200–203
32. Sharma J, Chhabra R, Andersen CS, Gothelf KV, Yan H, Liu Y (2008) J Am Chem Soc 130:7820–7821
33. Ding B, Deng Z, Yan H, Cabrini S, Zuckermann RN, Bokor J (2010) J Am Chem Soc 132:3248–3249
34. Zhao Z, Jacovetty E, Liu Y, Yang H (2011) Angew Chem Int Ed 50:2041–2044
35. Kuzuya A, Koshi N, Kimura M, Numajiri K, Yamazaki T, Ohnishi T, Okada F, Komiyama M (2010) Small 6:2664–2667
36. Endo M, Yang Y, Emura T, Hidaka K, Sugiyama H (2011) Chem Commun 47:10743–10745

37. Maune HT, Han S-P, Barish RD, Bockrath M, Goddard IIA, RothemundPaul WK, Winfree E (2010) Nat Nanotechnol 5:61–66
38. Youngblood B, Reich NO (2006) J Biol Chem 281:26821–26831
39. Endo M, Katsuda Y, Hidaka K, Sugiyama H (2010) J Am Chem Soc 132:1592–1597
40. Bruner SD, Norman DPG, Verdine GL (2000) Nature 403:859–866
41. Morikawa K, Matsumoto O, Tsujimoto M, Katayanagi K, Ariyoshi M, Doi T, Ikehara M, Inaoka T, Ohtsuka E (1992) Science 256:523–526
42. Endo M, Katsuda Y, Hidaka K, Sugiyama H (2010) Angew Chem Int Ed 49:9412–9416
43. Sannohe Y, Endo M, Katsuda Y, Hidaka K, Sugiyama H (2010) J Am Chem Soc 132:16311–16313
44. Xu Y, Sato H, Sannohe Y, Shinohara K-I, Sugiyama H (2008) J Am Chem Soc 130:16470–16471
45. Birkedal V, Dong M, Golas MM, Sander B, Andersen ES, Gothelf KV, Besenbacher F, Kjems J (2011) Microsc Res Techn 74:688–698
46. Steinhauer C, Jungmann R, Sobey TL, Simmel FC, Tinnefeld P (2009) Angew Chem Int Ed 48:8870–8873
47. Lin C, Jungmann R, Leifer AM, Li C, Levner D, Church GM, Shih WM, Yin P (2012) Nat Chem 4:832–839
48. Jungmann R, Steinhauer C, Scheible M, Kuzyk A, Tinnefeld P, Simmel FC (2010) Nano Lett 10:4756–4761
49. Yurke B, Turberfield AJ, Mills AP, Simmel FC, Neumann JL (2000) Nature 406:605–608
50. Bath J, Turberfield AJ (2007) Nat Nanotechnol 2:275–284
51. Yan H, Zhang X, Shen Z, Seeman NC (2002) Nature 415:62–65
52. Liao S, Seeman NC (2004) Science 306:2072–2074
53. Ding B, Seeman NC (2006) Science 314:1583–1585
54. Gu H, Chao J, Xiao S-J, Seeman NC (2009) Nat Nanotechnol 4:245–248
55. Gu H, Chao J, Xiao S-J, Seeman NC (2010) Nature 465:202–205
56. Lund K, Manzo AJ, Dabby N, Michelotti N, Johnson-Buck A, Nangreave J, Taylor S, Pei R, Stojanovic MN, Walter NG, Winfree E, Yan H (2010) Nature 465:206–210
57. Wickham SFJ, Endo M, Katsuda Y, Hidaka K, Bath J, Sugiyama H, Turberfield AJ (2011) Nat Nanotechnol 6:166–169
58. Bath J, Green SJ, Turberfield AJ (2005) Angew Chem Int Ed 44:4358–4361
59. Wickham SFJ, Bath J, Katsuda Y, Endo M, Hidaka K, Sugiyama H, Turberfield AJ (2012) Nat Nanotechnol 7:169–173
60. Castro CE, Kilchherr F, Kim D-N, Shiao EL, Wauer T, Wortmann P, Bathe M, Dietz H (2011) Nat Meth 8:221–229
61. Mei Q, Wei X, Su F, Liu Y, Youngbull C, Johnson R, Lindsay S, Yan H, Meldrum D (2011) Nano Lett 11:1477–1482
62. Vollmer J, Krieg AM (2009) Adv Drug Delivery Rev 61:195–204
63. Latz E, Verma A, Visintin A, Gong M, Sirois CM, Klein DCG, Monks BG, McKnight CJ, Lamphier MS, Duprex WP, Espevik T, Golenbock DT (2007) Nat Immunol 8:772–779
64. Li J, Pei H, Zhu B, Liang L, Wei M, He Y, Chen N, Li D, Huang Q, Fan C (2011) ACS Nano 5:8783–8789
65. Mohri K, Nishikawa M, Takahashi N, Shiomi T, Matsuoka N, Ogawa K, Endo M, Hidaka K, Sugiyama H, Takahashi Y, Takakura Y (2012) ACS Nano 6:5931–5940
66. Schüller VJ, Heidegger S, Sandholzer N, Nickels PC, Suhartha NA, Endres S, Bourquin C, Liedl T (2011) ACS Nano 5:9696–9702
67. Douglas SM, Bachelet I, Church GM (2012) Science 335:831–834

Chapter 2
Direct Observation of Single Hybridization and Dissociation of Photoresponsive Oligonucleotides in the Designed DNA Nanostructure

2.1 Introduction

Hybridization is the process that two complementary single-stranded nucleic acid molecules (DNA or RNA) associate together into a single double-stranded helix through specific base pairing. This process is reversible and the duplex can be controlled to dissociate by regulating the reaction conditions. To direct observation of the hybridization/dissociation reaction at single molecule level is still a challenge and important issue for the basic nanoscience and nanotechnology. Single molecules exhibit different characteristic, which is difficult to identify one by one from the ensemble state in bulk conditions. Individual molecules can behave heterogeneously, depending on the physical state of the molecules and the local environment around them [1–3].

DNA nanostructures have already been employed as scaffolds for enzyme reaction analysis and also the visualization of chemical reactions using various analysis methods [4–9]. Direct visualization of the association and dissociation of DNA duplex would also be quite fascinating. However, currently there is no direct and viable method to realize it under physical conditions. Our group have developed a single-molecule observation platform for the investigation of DNA–protein and structural DNA strand interactions in real-time using high-speed atomic force microscopy (AFM) [10–13]. In this chapter, a photocontrolled DNA nanostructure was constructed. And the dynamic hybridization and dissociation behaviors of a pair of pesudocomplementary photoresponsive oligonucleotides were tried to observe on this DNA nanostructure at single molecule level.

Azobenzene is one of photoresponsive molecules and gained popular attention of researchers because of its most intriguing property: reversible photoisomerization in a wavelength-dependent manner. Planer trans-form azobenzene molecules can isomerize to the non-planer cis-form by the photoirradiation of UV light (330 nm $\leq \lambda \leq 380$ nm) and the reverse transformation can easily switched by irradiation with visible light ($\lambda \geq 400$ nm) [14]. Asanuma and his collogues have

© Springer Japan 2015
Y. Yang, *Artificially Controllable Nanodevices Constructed by DNA
Origami Technology*, Springer Theses, DOI 10.1007/978-4-431-55769-2_2

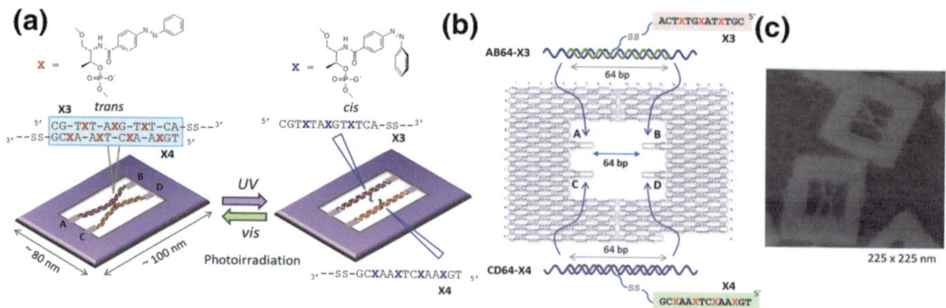

Fig. 2.1 Single-molecule observation system for the hybridization and dissociation of a photoresponsive DNA duplex containing azobenzene molecules. **a** Two dsDNAs carrying photoresponsive domains were assembled into a frame structure in X-shape because the photoresponsive domain hybridized to form a duplex in the *trans*-form of azobenzene moities. After UV irradiation, the photoresponsive domain dissociated because of the photoisomerization from trans- to cis- form, resulting the two dsDNAs in separated-shape. **b** Two dsDNAs containing photoresponsive domain (*X3* and *X4*) separately were located at specific sites (*A–B* and *C–D*) in the DNA frame using the corresponding ssDNAs as sticky ends. **c** AFM image of the assembled structure. [Single-Molecule Visualization of the Hybridization and Dissociation of Photoresponsive Oligonucleotides and Their Reversible Switching Behavior in a DNA Nanostructure/Endo, Masayuki; Yang, Yangyang; Suzuki, Yuki; Hidaka, Kumi; Sugiyama, Hiroshi/Angew. Chem. Int. Ed, 51/42. Copyright (c) [2012] [copyright owner as specified in the Journal]. (http://onlinelibrary.wiley.com/doi/10.1002/anie.201205247/abstract)

developed a series of azobenzene-tethered oligonucleotides, in which the photoresponsive performances were investigated comprehensively and also some further applications were successfully carried out [14–16]. It has already been found out that azobenzene molecules can reversibly regulate the hybridization of a double-stranded DNA by switching the photoirradiation. However, the dynamic process of single conformational changes of DNA strand has not been fully addressed yet.

A pair of pesudocomplementary DNA strands containing azobenzene moities (photoresponsive domains) was introduced into a DNA nanoframe that has already been employed to image DNA–enzyme interactions and observe G-quadruplex formation. As shown in Fig. 2.1, the photoresponsive domain is consist of two pseudocomplementary oligonucleotides in which one contains three azobenzenes (**X3**) while the other is tethered with four azobenzenes (**X4**). They can form the duplex when azobenzene molecules are in *trans*-form, which will be induced to dissociate in the *cis*-form after the UV irradiation. Two double-stranded DNAs (dsDNAs) carrying photoresponsive domain were assembled into the cavity (approximately 40 nm × 40 nm) of DNA nanoframe. The reversible switching performances of photoresponsive domain can be identified by a global structural change of two dsDNA as an X-shape and as a separated-shape in the DNA nanoframe, respectively.

2.2 Experimental Section

2.2.1 Synthesis of Oligonuleotides Containing Photoresponsive Domains

The single DNA strand "AB64-3-SS" (32 m, $\sim$100 μM, 20 μL) tethered with disulfide bond at the 5' terminal were mixed with 5 mM DTT and 0.1 M Tris–HCl buffer (pH 8.0). After mixing uniformly, the mixture was incubated at 37 °C for 2 h. The thiol-attached oligonucleotide (ODN-1) was purified by HPLC [linear gradient using 2–30 % acetonitrile/water (20 min) containing 20 mM ammonium formate, Nacalai Cosmosil C18 reversed-phase column (7.5 × 150 mm), 2.0 mL/min, 260 nm]. After the purification, the reduced thiol-attached oligonucleotide was directly treated with 5, 5'-dithiobis (2-nitrobenzoic acid) (DTNB) (1 μL, 50 mM DMF solution) at 40 °C for 1 h during the evaporation of HPLC solvent. The oligonucleotide–nitrobenzoic acid conjugate (ODN-2) was purified by HPLC [the same conditions as described above] (see Fig. 2.2).

The azobenzene tethered oligonucleotide ("Azo-3X- SS", 40 μM, 20 μL), which also had the disulfide protection at the 3' terminal, was also treated with 5 mM DTT and 0.1 M Tris–HCl buffer (pH 8.0) at 37 °C for 2 h for the de-protection. The thiol-attached oligonucleotide (ODN-3) was purified by HPLC [the same conditions as described above].

Finally, the same equivalent ODN-2 and ODN-3 were mixed and put at 40 °C for 1 h under the evaporation of HPLC solvent. The concentrated mixture was analyzed and purified by HPLC [the same conditions as described above] and finally lyophilized to get the azobenzene modified oligonucleotide (AB64-X3). The final product was confirmed by denaturing polyacrylamide gel electrophoresis. The method for the coupling of another strand CD64-3-SS with Azo-4X-SS was the same as the AB64-X3.

2.2.2 Preparation of the DNA Frame (Fig. 2.1b)

The DNA frame structure was prepared according to the previous method. Briefly, the DNA frame was assembled in a 20 μL solution containing 10 nM M13mp18 single-stranded DNA (New England Biolabs), 50 nM staple strands (5 eq), 20 mM Tris buffer (pH 7.6), 1 mM EDTA, and 10 mM $MgCl_2$ as following the previous study. The mixture was annealed from 85 to 15 °C at a rate of -1.0 °C/min.

2.2.3 Introduction of Two dsDNAs Containing Photoresponsive Domains into the DNA Frame

The pre-assembled dsDNAs containing photo-responsive domains [20 nM (two equivalents)] were incorporate into the DNA frame (10 nM) by heating at 40 °C and then cooling to 15 °C at a rate of −0.5 °C/min using a thermal cycler. The sample was purified using gel-filtration (GE sephacryl-300). The assembled structures were observed by AFM, and the yield of incorporation was counted manually.

2.2.4   Photoirradiation to the dsDNA-Attached DNA Frame

Photoirradiation for the sample was performed using Xe-lamp (300 W, Ashahi-spectra MAX-303) with band-path filter (10 nm—HW); 350 and 450 nm for the UV-light and the visible-light, respectively. The sample containing 10 nM dsDNA-attached DNA frame, 20 mM Tris–HCl (pH 7.6), 10 mM $MgCl_2$ was irradiated at 25 °C for 15 min for both UV- and visible-light.

2.2.5   High-Speed AFM Imaging of the dsDNAs in the DNA Frame

High-speed AFM images were obtained on a fast-scanning AFM (Nano Live Vision, RIBM, Tsukuba, Japan) using a silicon nitride cantilever (Olympus BL-AC10EGS) cooperated with light source (Olympus U-RFL-T). The sample (2 μL) was absorbed on a freshly cleaved mica plate for 5 min at room temperature, and then washed with the buffer solution for the observation of single photo-switching directly. Scanning was performed in the same buffer solution using a tapping mode.

2.3   Results and Discussion

The preparation of photoresponsive DNA nanoframe followed two steps, including (1) assembly of DNA frame as observation scaffold and (2) introduction of two dsDNAs carrying azobenzene moieties into the nanostructure. The assembled structures were confirmed by AFM (Fig. 2.1b, c). Two dsDNAs were located in the DNA nanoframe and connected at the center by the hybridization of the photoresponsive domain in X-shape.

2.3.1   Direct Imaging of the Single Photoresponsive Duplex by AFM

Figure 2.3 illustrates the molecular modeling of azobenzene modified DNA duplex. The azobenzene molecules in *trans*-form were intercalated into the duplex.

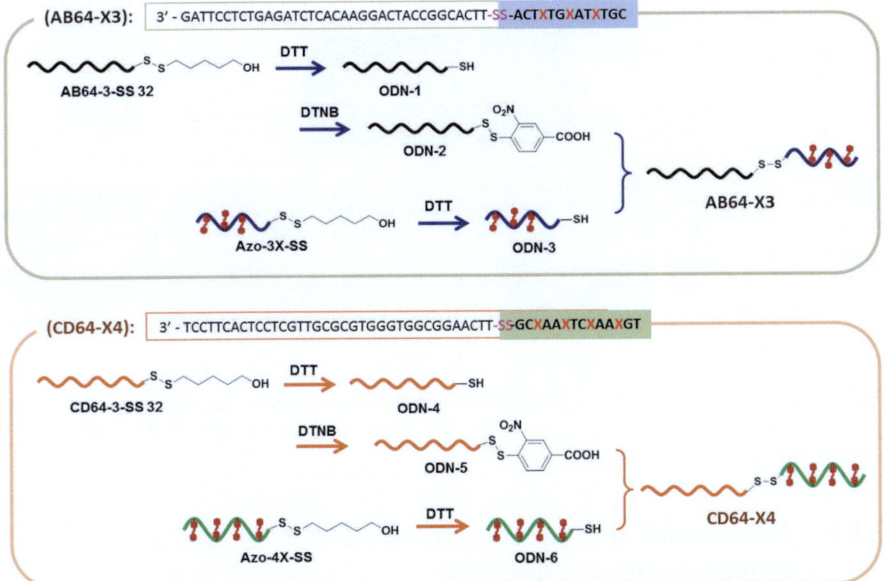

Fig. 2.2 Schematic drawings for oligonucleotides containing photoresponsive domains. The *red X* and *red hexagonal* pairs are defined as azobenzene molecules tethered with oligonucleotides. [Single-Molecule Visualization of the Hybridization and Dissociation of Photoresponsive Oligonucleotides and Their Reversible Switching Behavior in a DNA Nanostructure/Endo, Masayuki; Yang, Yangyang; Suzuki, Yuki; Hidaka, Kumi; Sugiyama, Hiroshi/Angew. Chem. Int. Ed, 51/42. Copyright (c) [2012] [copyright owner as specified in the Journal]. (http://onlinelibrary. wiley.com/doi/10.1002/anie.201205247/abstract)

The total length of the photoresponsive domains in duplex formation was estimated to be 10.5 nm including 17 base pairs (5.8 nm) and two organic linkers (4.7 nm). Two double stranded DNA (dsDNAs) were both 64-m ($\sim$20nm), which is fitted into the frame cavity in a tensed state with four connectors (**A–B**) and (**C–D**). Ideally they were placed parallelly in a distance of 15 nm before introducing photoresponsive domains.

After the two dsDNAs being coupled with photoresponsive domains separately, the hybridized photoresponsive domain was supported by tensed dsDNAs in the middle point as the X-shape, meaning that the photorcsponsive domain was also in a relatively tensed state imaged directly by AFM. Four representative AFM images of DNA nanoframe were shown in Fig. 2.3, in which the single photoresponsive duplex on each structure was clearly identified between two dsDNAs. And also the observed lengths of hybridized photoresponsive duplex between two dsDNAs were very close to the estimated length.

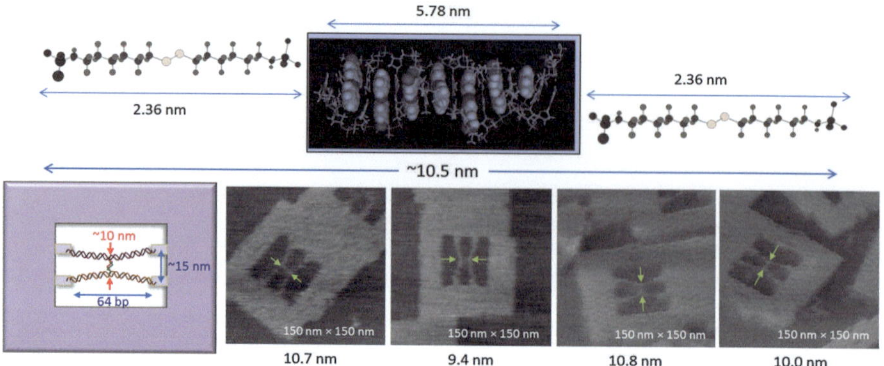

Fig. 2.3 Molecular modeling of azobenzene-tethered photoresponsive domains and AFM images of final assembled nanostructures in which the single photoresponsive duplex were clearly identified in each nanostructure

2.3.2 Photoirradiation of Oligonucleotides Containing Photoresponsive Domains

The photoisomerization performances of the photoresponsive domains were firstly examined by using two light sources that belong to (1) Xe-lamp and (2) high-speed AFM. The UV–visible spectroscopy was carried out for both UV and visible light irradiation under different time. The single-stranded strand, Azo-4X-SS, was chosen for the characterization. Firstly, the photoirradiation was carried out by using the Xe-lamp (Asahi-spectra) (Fig. 2.4a, b). Under UV irradiation (350 nm by band path filter) with different time, the absorbance of azobenzene molecules at 330 nm reduced gradually and a small increase at 440 nm occurred. After switching to visible light (450 nm by band path filter), the absorbance changed reversibly. The reversible absorbance change of azobenzene molecules indicated that Xe-lamp is able to initiate the photoisomerization of azobenzenes between *trans-* and *cis-*form. Next photoirradiation to oligonucleotide was also investigated by using another light source which is equipped together with high-speed AFM for single molecule analysis. Similar changes of the absorbance of azobenzenes were observed on UV–Vis spectras as expectedly (Fig. 2.4c, d). As a result, both of the light sources can successfully induce the photoreaction of photoresponsive domain.

2.3.3 Photoirradiation of Photoresponsive Nanoframe in Solution

The final assembled nanoframes were confirmed by AFM and ∼80 % of X-shaped DNA nanoframes were obtained because of the hybridization of photoresponsive domain containing azobenzene moieties in the *trans*-form (Fig. 2.5a). UV-light irradiation ($\lambda = 350$ nm by band path filter) from Xe-lamp was firstly introduced to nanoframe solution for 15 min at 25 °C. The irradiated sample was then imaged by AFM to confirm the structural change directly, by which separated-shaped nano-frames were observed and the distribution ratio increased from 16 to 26 % (Fig. 2.5b). The changes demonstrated that the hybridized photoresponsive domains were dissociated by the photoisomerization of azobenzenes from *trans*-form to *cis*-form. The UV-irradiated sample was further treated with visible-light ($\lambda = 450$ nm by band path filter) for 10 min at 25 °C. The ratio of X-shaped frame recovered from 64 to 74 % (Fig. 2.5c), representing that the reverse hybridization proceeded by the *cis*- to *trans*-isomerization. From the graph in Fig. 2.5d, it can be seen that the ratio variations of X-shape and separated-shape are consistent with each other during photoirradiation. The photoresponsive domain in DNA nanostructure worked correctly in solution in a wavelength-dependent manner.

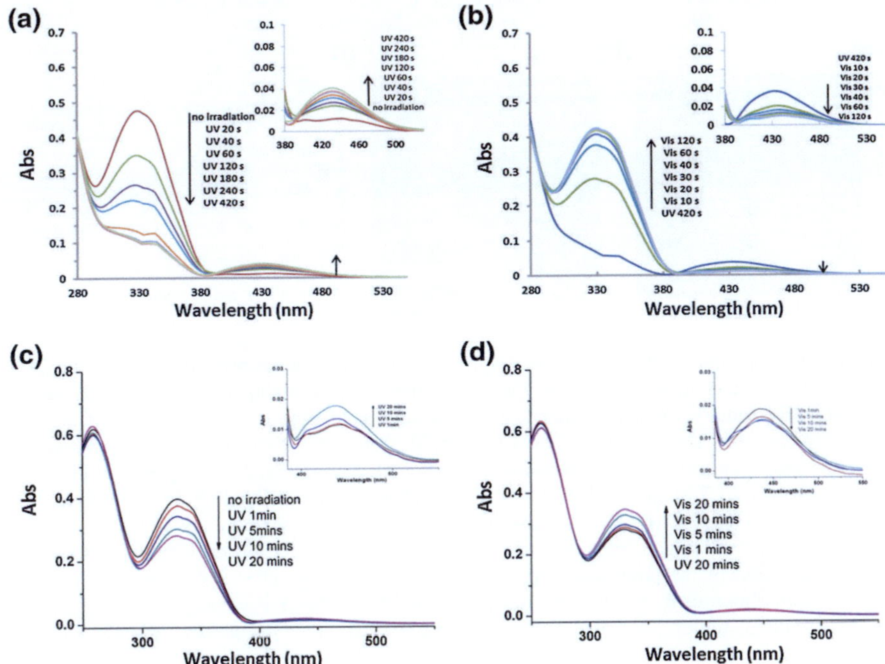

Fig. 2.4 UV–Vis spectra of unmodified Azo-4X-SS under different irradiation wavelength and irradiation time by using (1) light source of Xe-lamp (**a** and **b**) and (2) light source equipped on high-speed AFM (**c** and **d**). Conditions: 10 mM, pH 7.6 Tris–HCl buffer, UV light (350 nm) and visible-light (450 nm) by band path filter, room tempterature. [Single-Molecule Visualization of the Hybridization and Dissociation of Photoresponsive Oligonucleotides and Their Reversible Switching Behavior in a DNA Nanostructure/Endo, Masayuki; Yang, Yangyang; Suzuki, Yuki; Hidaka, Kumi; Sugiyama, Hiroshi/Angew. Chem. Int. Ed, 51/42. Copyright (c) [2012] [copyright owner as specified in the Journal]. (http://onlinelibrary.wiley.com/doi/10.1002/anie.201205247/abstract)

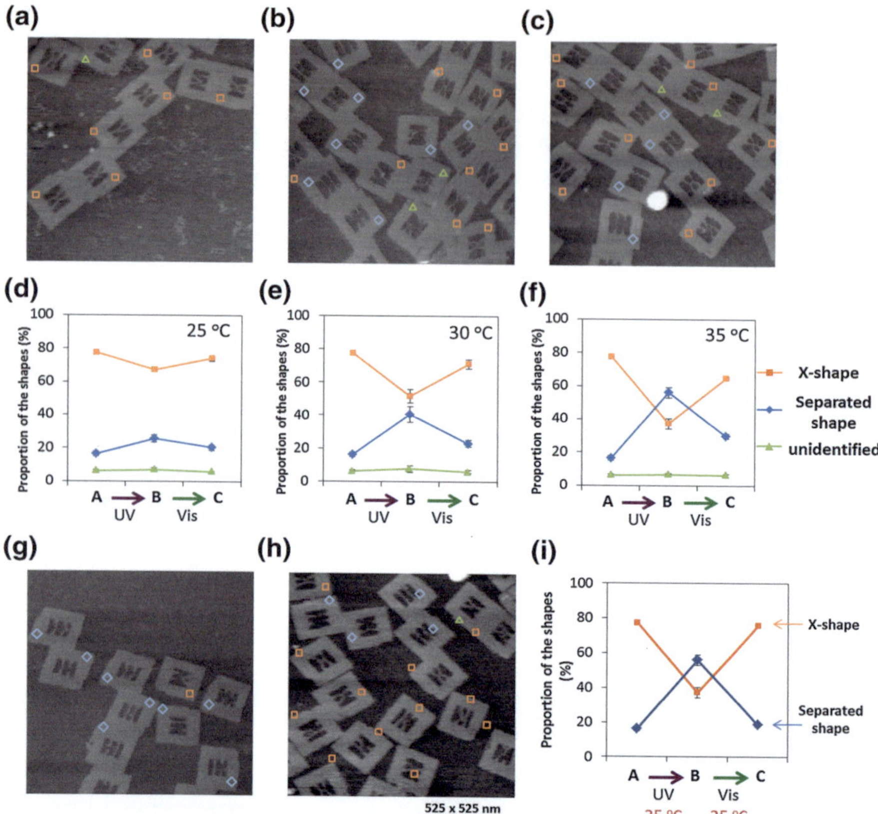

Fig. 2.5 Photoregulative dissociation and hybridization of photoresponsive domains in the DNA nanostructure. AFM images of DNA frames **a** after the assembling two dsDNAs containing pseudocomplementary strands; **b** after UV irradiation for 15 min at 25 °C; **c** after visible-light irradiation for 15 min at 25 °C of the sample from (**b**). X-shape = *orange rectangle*; separated-shape = *blue diamond*; unidentified shapes = *green triangle*. **d–f** Temperature-dependent distribution ratio of X-shapes and separated-shapes after UV and visible-light irradiation at 25 °C (**d**), 30 °C (**e**), and 35 °C (**f**). A = after assembly, B = after UV irradiation, and C = after UV then visible-light irradiation. **g, h** Temperature-controlled photoirradiation for the effective dissociation and hybridization of the photoresponsive domains. AFM images after UV irradiation at 35 °C (**g**) and with subsequent visible-light irradiation at 25 °C (**h**). **i** Percentage of X-shapes and separated-shapes under the predetermined photoirradiation temperatures. [Single-Molecule Visualization of the Hybridization and Dissociation of Photoresponsive Oligonucleotides and Their Reversible Switching Behavior in a DNA Nanostructure/Endo, Masayuki; Yang, Yangyang; Suzuki, Yuki; Hidaka, Kumi; Sugiyama, Hiroshi/Angew. Chem. Int. Ed, 51/42. Copyright (c) [2012] [copyright owner as specified in the Journal]. (http://onlinelibrary.wiley.com/doi/10.1002/anie.201205247/abstract)

Besides the wavelength, it has already been found out that reaction temperature is also the major factor for the photoregulative hybridization and association [16]. To evaluate the temperature effect on photoresponsive domain in DNA nanoframe during photoirradiation, three temperatures: 25, 30 and 35 °C were examined through one cycle of irradiation: UV-15 min and vis-10 min. The ratios of X-shape and separated-shape were analyzed, respectively (Fig. 2.5d–f). It can be clearly seen that higher reaction temperature exhibited higher conversion to separated-shape (56 % at 35 °C with UV light treatment), indicating that higher temperature favored the dissociation of photoresponsive domain under UV irradiation. However, it was observed that higher temperature weakened the re-formation of X-shape by the hybridization of photoresponsive domain with visible light, meaning that lower temperature is better for the hybridization of already dissociated domain. From these results, it was anticipated that controlled reaction temperature is necessary to obtain efficient dissociation and re-hybridization. (1) 35 °C for UV and (2) 25 °C for visible light were finally employed as reaction temperature for one-cycle photoirradiation. The AFM images of irradiated solution were shown in Fig. 2.5g, h respectively. From the percentage changes of X-shape and separated-shape shown in Fig. 2.5i, the UV-vis irradiated sample almost returned to the initial state completely under controlled temperature.

2.3.4 *Direct Observation of Dissociation of Photoresponsive Domains in Nanoframe Under UV Irradiation*

In bulk solution conditions, it was verified that this photoregulative DNA nanoframe is not only wavelength-dependent but also temperature-dependent. For single molecule analysis by high-speed AFM, constant temperature is preferable for observation. In this case, the AFM observation system was set up at 25 °C and run through whole observing process during photoirradiation, which is much lower than the melting temperature of the photoresponsive domain, to exclude the temperature effect on its dissociation.

Firstly, a freshly prepared solution of DNA nanoframe assembled with photoresponsive domain was loaded onto a clear mica surface. After clear AFM images of DNA nanoframe in X-shape were obtained, the UV light ($\lambda = 330$–380 nm) was switched by band path filter and directly introduced to the cantilever stage on high-speed AFM, which can initiate the *trans-* to *cis-*isomerization during the scanning. The time-lapsed AFM images representing the two dsDNAs changing from X-shape to separated-shape on nanoframe were shown in Fig. 2.6, indicating the dissociation of photoresponsive domain by direct UV irradiation. Interestingly, the dissociation of each single nanoframe occurred at different moments separately from 30 to 70 s. In single molecule resolution, single event of different single molecules were captured by high-speed AFM. Also, some dsDNAs on nanoframe did not show any response to photoirraidation, which may be attributed to the strong interactions between DNA molecules and mica surface, resulting the suppression of the movement of dsDNAs.

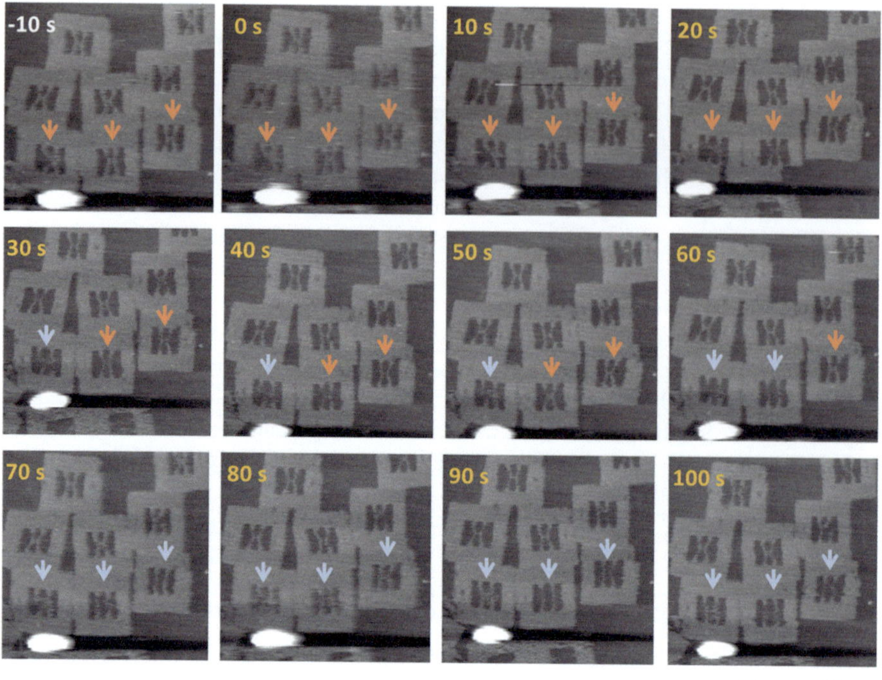

375 nm × 375 nm

Fig. 2.6 Dissociation of the photoresponsive domain in the DNA frame during UV irradiation. AFM scanning was 0.1 frames per second. X-shape = *orange arrow*; separated-shape = *blue arrow*. [Single-Molecule Visualization of the Hybridization and Dissociation of Photoresponsive Oligonucleotides and Their Reversible Switching Behavior in a DNA Nanostructure/Endo, Masayuki; Yang, Yangyang; Suzuki, Yuki; Hidaka, Kumi; Sugiyama, Hiroshi/Angew. Chem. Int. Ed, 51/42. Copyright (c) [2012] [copyright owner as specified in the Journal]. (http://onlinelibrary. wiley.com/doi/10.1002/anie.201205247/abstract)

2.3.5 *Direct Observation of Association of Photoresponsive Domains in Nanoframe Under Visible Light Irradiation*

Under UV irradiation, the dissociation behaviors of single pair of photoresponsive domain on DNA nanoframe were directly visualized in real-time. Next it is tried to observe the reverse hybridization movements with the irradiation of visible light. The sample pretreated with UV light was loaded on the mica surface and the expected nanoframes in separated-shape were imaged. Visible light (λ = 440–470 nm) was then introduced by band path filter for the *cis*- to *trans*-isomerization of azobenzene molecules in the nanostructures. As shown in Fig. 2.7, the two separated dsDNAs were associated at the central position into X-shape because of the hybridization of photoresponsive domain (100–120 s). And also the

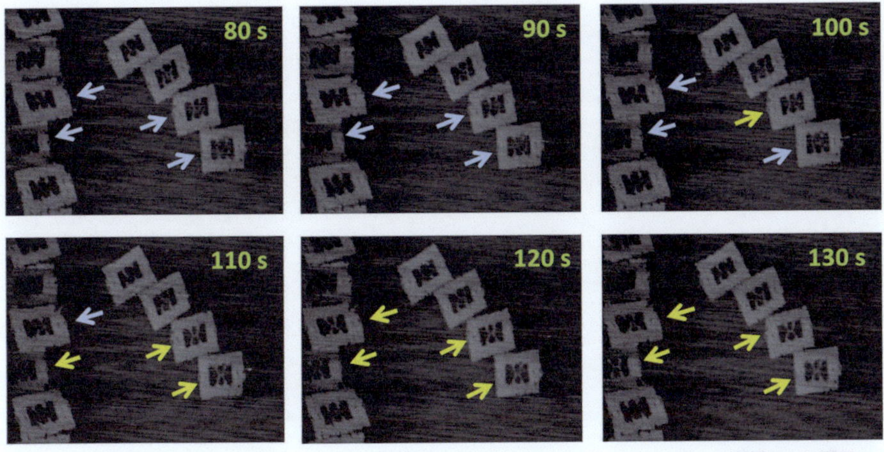

Fig. 2.7 Hybridization of the pre-UV irradiated photoresponsive domain in the DNA frame during visible-light irradiation. AFM scanning rate was 0.1 frames per second. X-shape = *yellow arrow*; separated-shape = *blue arrow*. [Single-Molecule Visualization of the Hybridization and Dissociation of Photoresponsive Oligonucleotides and Their Reversible Switching Behavior in a DNA Nanostructure/Endo, Masayuki; Yang, Yangyang; Suzuki, Yuki; Hidaka, Kumi; Sugiyama, Hiroshi/Angew. Chem. Int. Ed, 51/42. Copyright (c) [2012] [copyright owner as specified in the Journal]. (http://onlinelibrary.wiley.com/doi/10.1002/anie.201205247/abstract)

hybridization on different nanoframes occurred separately at different time points when observing the events at single molecule resolution (Fig. 2.8).

The photoisomerization of azobenzene as well as its derivatives is extremely rapid, occurring on picosecond timescales[17], which can be ignored as compared with the reaction time-scale for the hybridization and dissociation of the azobenzene-modified oligonucleotides. Therefore, the rate determining process of dynamic interacting between two dsDNAs should be the hybridization and dissociation of photoresponsive domains. The first UV-induced dissociation occurred at 30 s while the first re-hybridization was captured at 110 s, representing that the slower hybridization of two single DNA strands requires more time for the correct contact and complete association.

2.3.6   Reversible Photoswitching of Photoresponsive Domains Repeatly in a Single DNA Nanoframe

Besides the direct visualization of hybridization and dissociation of photoresponsive domains on DNA nanoframe separately on mica surface, the reversible activities of a single pair were also successfully obtained by alternating UV–Vis–UV directly on the AFM stage. The X-shaped and separated-shaped DNA

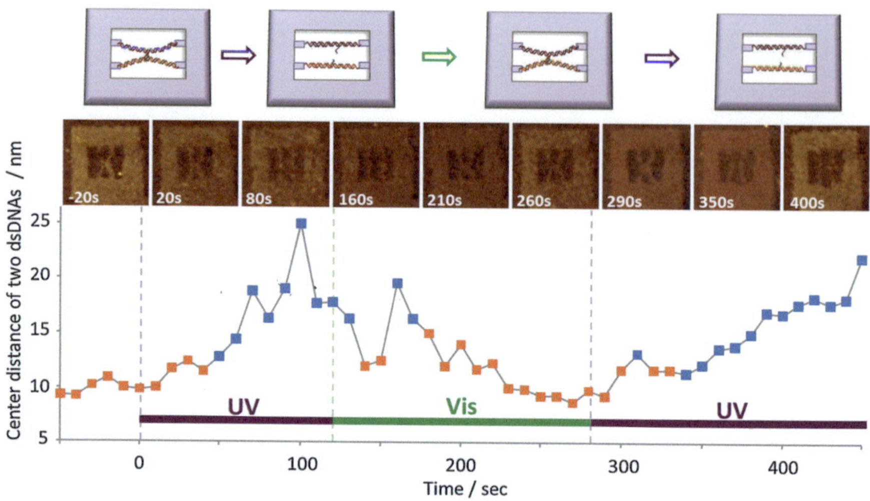

Fig. 2.8 Photoswitching activities of single pair of photoresponsive domain in the DNA frame. The closest distances between two dsDNAs were plotted by alternating photoirradiation in UV (*purple*) and visible-light range (*green*) during AFM scanning. The dissociation and hybridization of photoresponsive domains were monitored repeatedly. [Single-Molecule Visualization of the Hybridization and Dissociation of Photoresponsive Oligonucleotides and Their Reversible Switching Behavior in a DNA Nanostructure/Endo, Masayuki; Yang, Yangyang; Suzuki, Yuki; Hidaka, Kumi; Sugiyama, Hiroshi/Angew. Chem. Int. Ed, 51/42. Copyright (c) [2012] [copyright owner as specified in the Journal]. (http://onlinelibrary.wiley.com/doi/10.1002/anie.201205247/abstract)

nanoframe could be repeated observed by high-speed AFM. As shown in Fig. 2.8, here the closest distance between two dsDNAs was analyzed to determine the state of photoresponsive domains. The first dissociation was occurred at 50 s after introducing UV-irradiation and after that the two dsDNAs maintained separated shape for over 90 s. After irradiation switching from UV to visible irradiation at 120 s, a quick association was captured at 140–150 s but dissociated at the next 10 s. it was assumed that this was occurred randomly such as incomplete hybridization. The photoresponsive domain further associated stably at 180 s, indicating that one cycle of the reversible photoswitching was accomplished. Next it took 60 s (at 340 s) to dissociate again at the second round of UV irradiation. As a result, the reversible dynamic behaviors of photoresponsive domain were successfully visualized continuously from the global structural changes of dsDNAs.

Normally photo-damage is one of unavoidable issues for the photoirradiation especially in UV range. Here, the appearances of both DNA nanoframe and the incorporated dsDNAs in nanoframe should be maintained intact through all irradiation processes in solution as well as on the mica surface. The photoirradiation conditions employed in this study should be moderate to maintain DNA safely.

2.4 Conclusion

Based on DNA origami technique a single molecule observation nanosystem are demonstrated here to observe the hybridization and dissociation of photoresponsive oligonucleotides directly by using high-speed AFM. The dynamic behaviors of photoresponsive duplex containing azobenzene molecules were captured by recording the movement and global change of two dsDNAs in the DNA nanoframe. By switching UV and visible-light irradiation, the dissociation and hybridization of the photoresponsive domain occurred reversibly in solution as well as on flatten mica surface. DNA-based nanoscaffolds can be employed as promising observation plateform for other kinds of switching behaviors in nanometer-sized resolution and also in subsecond timescale by using high-speed AFM.

References

1. Cornish PV, Ha T (2007) ACS Chem Biol 2:53–61
2. Hilario J, Kowalczykowski SC (2010) Curr Opin Chem Biol 14:15–22
3. Novo C, Funston AM, Mulvaney P (2008) Nat Nanotechnol 3:598–602
4. Ke Y, Lindsay S, Chang Y, Liu Y, Yan H (2008) Science 319:180–183
5. Voigt NV, Torring T, Rotaru A, Jacobsen MF, Ravnsbaek JB, Subramani R, Mamdouh W, Kjems J, Mokhir A, Besenbacher F, Gothelf KV (2010) Nat Nanotechnol 5:200–203
6. Yoshidome T, Endo M, Kashiwazaki G, Hidaka K, Bando T, Sugiyama H (2012) J Am Chem Soc 134:4654–4660
7. Nakata E, Liew FF, Uwatoko C, Kiyonaka S, Mori Y, Katsuda Y, Endo M, Sugiyama H, Morii T (2012) Angew Chem Int Ed 51:2421–2424
8. Rajendran A, Endo M, Sugiyama H (2012) Angew Chem Int Ed 51:874–890
9. Torring T, Voigt NV, Nangreave J, Yan H, Gothelf KV (2011) Chem Soc Rev 40:5636–5646
10. Endo M, Katsuda Y, Hidaka K, Sugiyama H (2010) J Am Chem Soc 132:1592–1597
11. Endo M, Katsuda Y, Hidaka K, Sugiyama H (2010) Angew Chem Int Ed 49:9412–9416
12. Sannohe Y, Endo M, Katsuda Y, Hidaka K, Sugiyama H (2010) J Am Chem Soc 132:16311–16313
13. Endo M, Tatsumi K, Terushima K, Katsuda Y, Hidaka K, Harada Y, Sugiyama H (2012) Angew Chem Int Ed 51:8778–8782
14. Asanuma H, Liang X, Nishioka H, Matsunaga D, Liu M, Komiyama M (2007) Nat Protoc 2:203–212
15. Liang, X, Takenaka, N, Nishioka, H, Asanuma, H (2008) Chem Asian J 3:553–560
16. Liang X, Mochizuki T, Asanuma H (2009) Small 5:1761–1768
17. Tamai N, Miyasaka H Chem Rev **2000**, 100, 1875–1890

Chapter 3
Direct Observation of Logic-Gated Dual-Switching Behaviors Inducing the State Transition in a DNA Nanostructure

3.1 Introduction

Not only acting as genetic carriers, DNA can also serve as basic building blocks for biomaterials, which has gained increasing interest in recent years [1, 2]. In particular, various types of DNA based circuits with different combinational inputs have been developed for logical operations [3–9] or processing finite-state programs [10–13]. Currently, the standard DNA finite-state automaton often consists of free-floating DNA strands, whose states can be switched by introduction of certain triggers, such as aptamer-substrate complexes and pH [14]. The bulk DNA state change can then be read out as a quantitative and indirect output from fluorescence or gel electrophoresis [12]. Due to the detection limitation of fluorescence and gel electrophoresis of free-floating state of DNA strands, it is often hard to achieve single molecule level resolution and real-time detection in the standard DNA finite-state automaton system. DNA origami allows the tethering of target single DNA strand in combination with high speed atomic force microscopy (AFM), which would enable real-time monitoring the state change of DNA strand at single molecule level [15]. This technique provides a viable tool to decipher the DNA strand transformation processes during state transition. Previously, our lab have developed a series of frame-shaped DNA origami tethered with single DNA strands functionalized with DNA motifs, such as G-telomeric repeats, [16] photoresponsive oligonucleotides, [17] and B-Z DNA transiting strands [18]. Such DNA nanostructures have been proved to be able to observe nanomechanical movements in response to one specific trigger using high-speed AFM at single molecule level and in real-time fashion [19]. In this study, we aim at incorporating two triggers and testing dual-switching behaviors between two different DNA motifs in a single nanoframe.

A nanoframe observation system was designed to integrate photoresponsive domains and G-telomeric repeats together. It was attempted to visualize logical association/dissociation interactions among three parallel double-stranded DNAs (dsDNAs) analogous to finite-state-transition in real time, shown in Fig. 3.1.

© Springer Japan 2015
Y. Yang, *Artificially Controllable Nanodevices Constructed by DNA Origami Technology*, Springer Theses, DOI 10.1007/978-4-431-55769-2_3

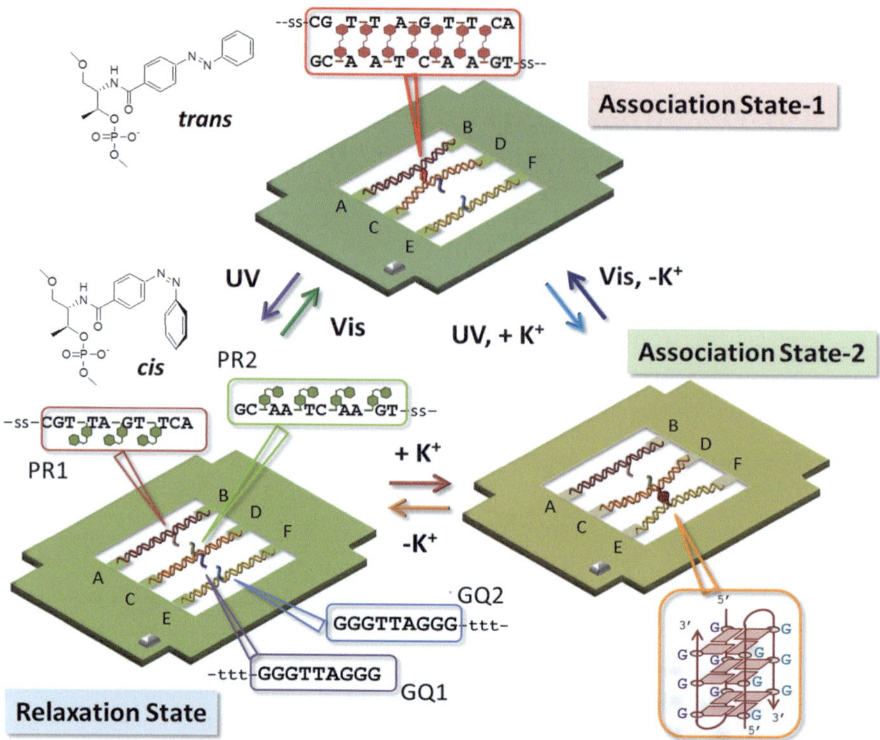

Fig. 3.1 Schematic presentation of the dual-switching behaviors of three parallel dsDNAs in our frame-shaped DNA origami system. There are three dsDNA configurations in a single nanoframe system, representing Association State-1 (AS-1), Relaxation State (RS) and Association State-2 (AS-2), respectively. Each state is triggered by combination of photo irradiation, and potassium ion. [Yang, Y.; Endo, M.; Suzuki, Y.; Hidaka, K.; Sugiyma, H. *Chem. Comm.*, **2013**, 50, 4211–4213.] Reproduced by permission of The Royal Society of Chemistry

A frame-shaped nanostructure had a vacant area inside, in which three pairs of connection sites (A–B, C–D, and E–F) were introduced for hybridization of the three different (parallel) dsDNAs (Figs. 3.1 and 3.2a). Two dsDNAs containing different azobenzene-modified (pseudocomplementary) photoresponsive oligonucleotides (PR-1 and PR-2) were hybridized to the A–B and C–D sites, respectively. The photoresponsive domains can hybridize in the *trans*-form or dissociate in the *cis*-form of azobenzene by photoirradiation with different wavelengths [20–22]. Meanwhile, two G-telomeric overhangs: 5′-TTTGGGTTAGGG-3′ (GQ-1) and 5′-GGGTTAGGGTTT-3′ (GQ-2) were introduced to the CD and EF strands, respectively, which rendered the capability of G-quadruplex formation between CD and EF strands [23]. The "kissing" and "unkissing" between AB and CD strands at the central positions, representing Association State-1 (AS-1) and Relaxation State

(RS) separately, can be turned on and off by photoirradiation at different wavelengths (UV vs. Vis). The association and dissociation between CD and EF via formation and disruption of G-quadruplex, can be regulated by addition and removal of potassium ion, resulting in RS and Association State-2 (AS-2) respectively. In this work, we have investigated the sequential state transformation from AS-1 (photoinduced dissociation) to AS-2 (G-quadruplex formation) by addition of photoirradiation at UV range and K^+ simultaneously. A direct observation of the reversible switching events among three independent dsDNAs (i.e. AB, CD, and EF) in a logical manner was also performed.

3.2 Experimental Section

3.2.1 Design of DNA Nanoframe Containing Six Connecting Positions

The new-type DNA nanoframe was designed by using caDNAno software. A rectangular-shaped structure in a size of ~ 100 nm $\times$ ~ 90 nm, holding a square-shaped cavity inside for the cooperating of dsDNAs with six connecting positions: A–B, C–D and E–F. Three different dsDNAs carrying predetermined synthetic oligonucleotides modified by disulfide bonds (the method for the synthesizing can be seen in Chap. 2.) were all placed in parallel arrangements.

3.2.2 Assembly of Dual-Switching DNA Nanoframe Containing Photoresponsive Oligonucleotides and G-Tracts

The DNA nanoframe was assembled in a 20 µL solution containing 10 nM M13mp18 single-stranded DNA, 50 nM staple strands (5 eq), 20 mM Tris buffer (pH 7.6), 1 mM EDTA, and 10 mM $MgCl_2$ as following the previous study. The mixture was annealed from 85 to 15 °C at a rate of −0.5 °C/min. The pre-assembled dsDNAs containing photo-responsive oligonucleotides [20 nM (two equivalents)] and G-tracts were incorporated into the DNA nanoframe (10 nM) by heating at 40 °C and then cooling to 15 °C at a rate of −0.1 °C/min using a thermal cycler. The sample was purified using gel-filtration. The assembled structures were confirmed by AFM (Fig. 3.3b, c) after the removal of excessive dsDNAs and the yield of incorporation was evaluated (Fig. 3.3b). A hairpin marker pointed by yellow arrow in AFM images was introduced close to the "E" site in order to distinguish dsDNAs hybridized to the frame.

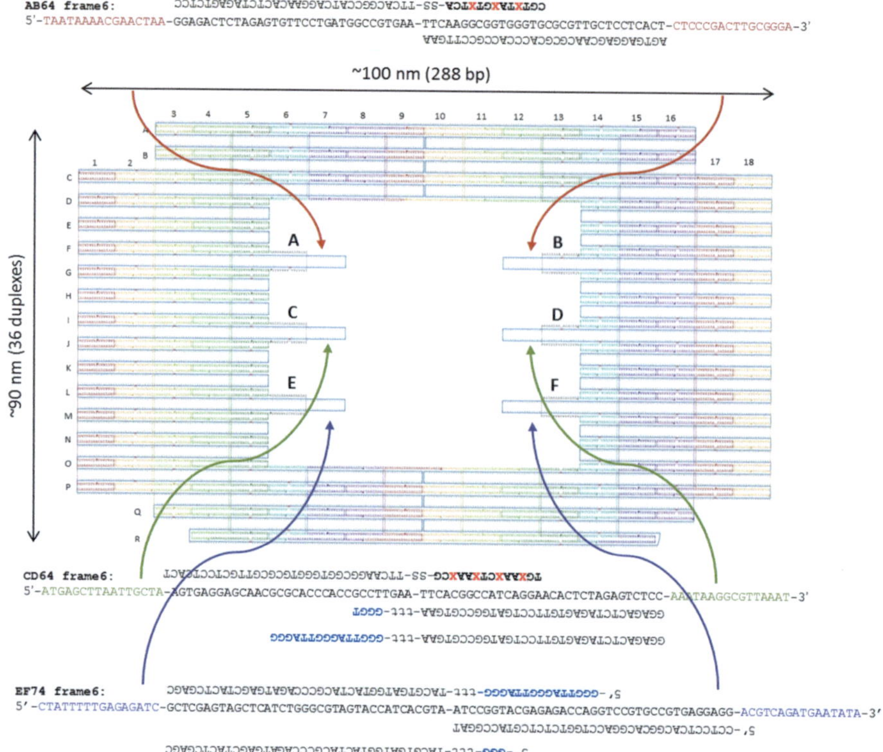

Fig. 3.2 Schematic design of DNA nanoframe containing six connecting positions (*A–B*, *C–D* and *E–F*) for the hybridization with three dsDNA carrying predetermined oligonuleotides. [Yang, Y.; Endo, M.; Suzuki, Y.; Hidaka, K.; Sugiyma, H. *Chem. Comm.*, **2013**, 50, 4211–4213.] Reproduced by permission of The Royal Society of Chemistry

3.2.3　Photoirradiation to the Dual-Switching DNA Nanoframe

Photoirradiation to the sample in tube was performed using Xe-lamp (300 W, Asahi-spectra MAX-303) with band-path filter (10 nm FWHM); 350 nm for UV-light and 450 nm for visible-light, respectively. The sample containing 10 nM DNA nanoframe, 20 mM Tris-HCl (pH 7.6), 10 mM MgCl$_2$ was irradiated at 35 °C for 10 min for UV-light.

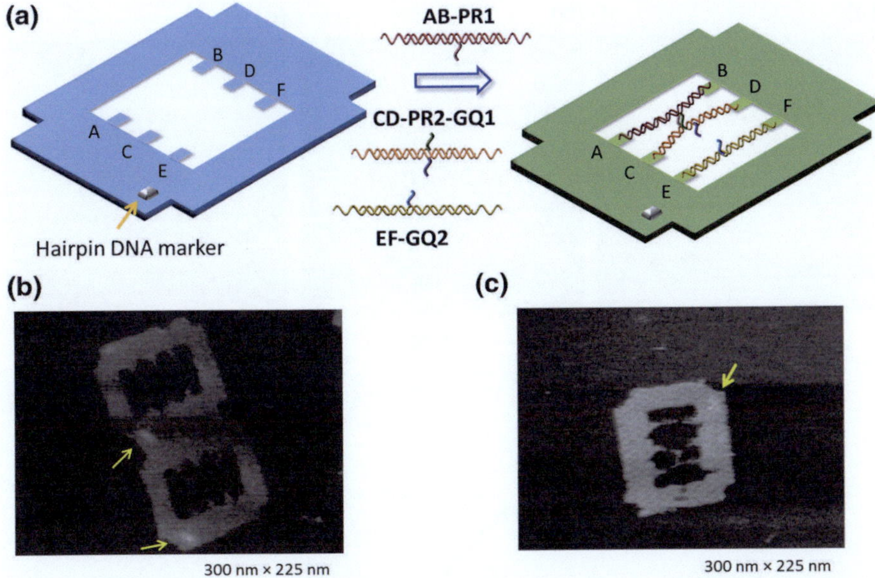

Fig. 3.3 a Brief structure of nanoframe with six connection sites for hybridization with three parallel dsDNAs and AFM images of DNA nanoframe **b** before and **c** after the introduction of three parallel dsDNAs carrying different DNA motifs in the central position. AB: photoresponsive oligonucleotides with three azobenzene molecules (PR-1); CD: photoresponsive oligonucleotides with four azobenzene molecules (PR-2) and two G-telomeric repeats sequence (GQ1); EF: two G-telomeric repeats (GQ2) sequence. A hairpin DNA marker was introduced close to "E" site pointed by *yellow arrow* in the AFM images (**b**) and (**c**). [Yang, Y.; Endo, M.; Suzuki, Y.; Hidaka, K.; Sugiyma, H. *Chem. Comm.*, **2013**, 50, 4211–4213.] Reproduced by permission of The Royal Society of Chemistry

3.3 Results and Discussion

3.3.1 Evaluation of the State Transition in Solution by Switching Photoirradiation and Potassium Ion

As shown in Fig. 3.4a–c, three patterns of nanostructures were mainly obtained in two kinds of association state (AS-1 and AS-2) and another kind of relaxation state (RS). For the primary sample (**switch-1**, without any triggering) the dominant product (approximately 62 %, Fig. 3.4d) was AS-1, in which AB and CD were associated because of the hybridization of photoresponsive motifs while the EF was left alone in a K^+-free buffer. However, it was noted that there were still 7 % of nanostructures exist in AS-2 state and 31 % of nanostructures exist in RS state. By adding K^+ (50 mM, **switch-2**) alone into the initial solution, the percentage of nanostructures at AS-1 state was slightly decreased while the percentage of both RS and AS-2 were slightly increased (Fig. 3.4d). This result suggests that the addition of K^+ alone had slight effect on all three states formation. On the other hand,

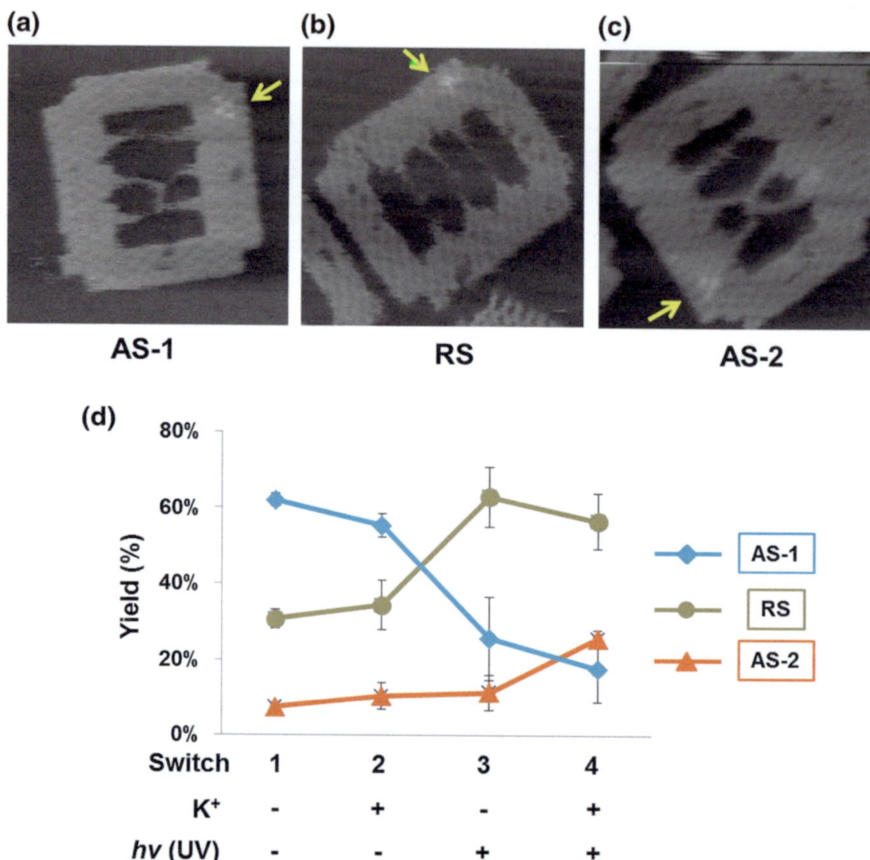

Fig. 3.4 AFM images of three finite states and the yield of three states under four independent inputs conditions. **Switch-1**: without any stimulation; **switch-2**: under buffer containing 50 mM K^+; **switch-3**: with UV irradiation for 10 min at 35 °C; **switch-4**: integrating 50 mM K^+ with UV irradiation for 10 min at 35 °C. The hairpin marker to distinguish three dsDNAs was pointed out by *yellow arrow*. *Bule diamond*: yield of AS-1; *tan circle*: yield of RS; *orange triangle*: yield of AS-2. The error bars represent that the experiments were all repeated three times. [Yang, Y.; Endo, M.; Suzuki, Y.; Hidaka, K.; Sugiyma, H. *Chem. Comm.*, **2013**, 50, 4211–4213.] Reproduced by permission of The Royal Society of Chemistry

by exposing to photoirradiation ($\lambda = 350$ nm band pass filter) for 10 min at 35 °C under the K^+-free buffer (**switch-3**), the percentage of AS-1 was decreased to 26 % while RS was increased to 63 % and AS-2 was slightly increased. By introducing two switches simultaneously (**switch-4**: adding K^+ and UV irradiation), the yield of AS-2 increased to 26 %, which is 3-fold higher than the yield in **switch-1** (Fig. 3.4d). The yields of three states changes under different combinations of triggers were summarized in the graph of Fig. 3.4d. It could be concluded that without the help of UV irradiation the associated photoresponsive domain inhibited the formation of G-quadruplex in the state of AS-2 regardless the presence of K^+. UV irradiation is necessary for high yield of RS. Regardless the presence of different switches, AS-2 state exists in small fraction of sample but the state transforming can be directly quantified by AFM imaging. In general, to some extent, it was able to achieve the cascading transformation from photoinduced dissociation to G-quadruplex formation in our nanostructures via UV irradiation and the addition of K^+.

3.3.2   Direct Visualization of the State Transition from AS-1 to AS-2 Using High-Speed AFM

Previously, it has already been successfully visualized the reversible association and dissociation between two parallel dsDNAs in a single nanostructure using K^+ [16] or photoirradiation [17] as switches. Here, these two external switches were integrated together to regulate three parallel dsDNAs behaviors in a logic-gated fashion using another nanoframe as observation platform, which performed similar as mechanical state transition. The reversible conversions from any three states to other ones were initiated and directly monitored by the combinations of switching stimuli using high-speed AFM. Similar as already reported results, the reversible conversions of AS-1/RS corresponding to photoinducing structural changes and AS-2/RS responsible for the ion-regulating structural changes were both successfully carried out in this new nanoscafflod by employing high-speed AFM (results now shown).

Next we tried to directly observe dissociation of AB and CD and sequential association between CD and EF by using UV irradiation in the presence of K^+ (10 mM) as dual-switches. The nanoframe sample was first prepared using K^+-free buffer and further loaded on a mica plate using same buffer. The fast scanning operation was carried out by changing the observation buffer containing 10 mM K^+ together with UV-irradiation (λ = 330–380 nm by band path filter) to induce the *trans*- to *cis*-form isomerization of azobenzene molecules of photoresponsive motifs. The AFM images were set up at a scanning rate of 0.2 frames per second. A nanoframe in the formation of AS-1 was focused for observation. As can be seen in Fig. 3.5, AB and CD were dissociated after the UV irradiation (65–70 s) owing to the dissociation of photoresponsive motifs in dsAB and dsCD. The dissociated CD immediately associated oppositely with the adjacent EF (70 s) because of the formation of G-quadruplex in the presence of K^+. The nanoframe was regulated to convert from "AS-1" to "AS-2" in a logical manner with combinations of two switches: UV-irradiation and K^+. Under the above observation conditions and stimuli of UV/K^+, the state changing from AS-1 to AS-2 was directly visualized, during which no other transition-state analogues were observed.

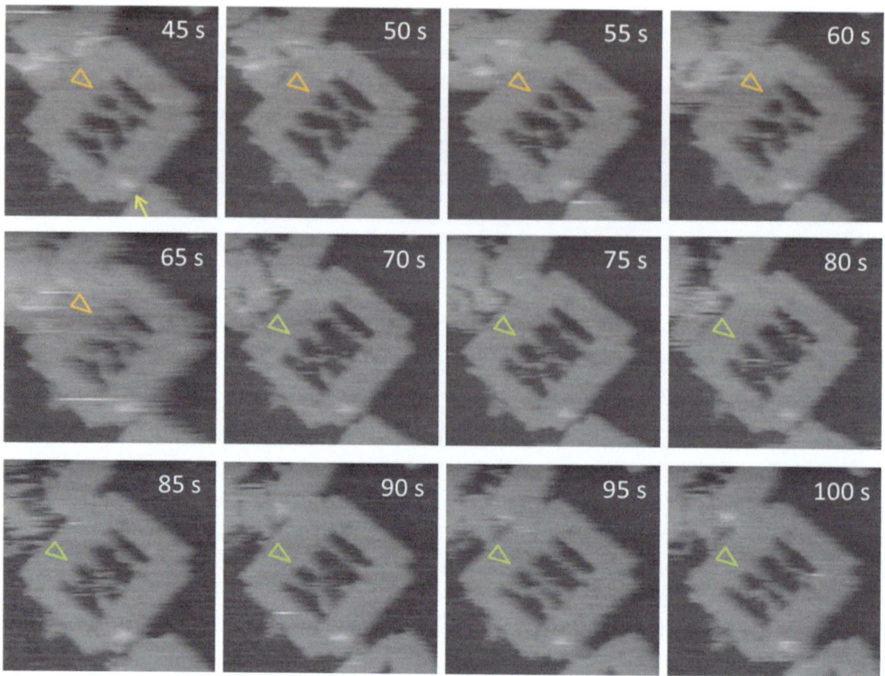

Fig. 3.5 Successive AFM images of the configuration change from AS-1 to AS-2 in a single nanoframe under UV photoirradiation in the observation buffer containing 10 mM K^+. AFM scanning rate was 0.2 frames per second. The lapsed time after starting photoirradiation was marked on each AFM image. The "kissing" between AB and CD was pointed by *green triangles* while the "kissing" between CD and EF was pointed by *orange triangles*. The hairpin marker was pointed by *yellow arrows*. Image size: 180 nm × 180 nm. [Yang, Y.; Endo, M.; Suzuki, Y.; Hidaka, K.; Sugiyma, H. *Chem. Comm.*, **2013**, 50, 4211–4213.] Reproduced by permission of The Royal Society of Chemistry

3.3.3 Direct Visualization of the Reverse State Transition from AS-2 to AS-1 Using High-Speed AFM

The reverse conversion from "AS-2" to "AS-1" was further examined by irradiating with visible light and under the observation buffer without K^+. A primary sample of nanoframe pre-treated with UV photoirradiation in the buffer containing K^+ was adsorbed on a mica surface and then washed sufficiently before fast scanning of AFM. By introducing K^+-free buffer for observation and alternating irradiation wavelength to visible light (λ = 440–470 nm) by band-pass filter, the AFM images were obtained under same scanning rate. The scanning results were shown in Fig. 3.6. At the beginning, a nanoframe in the formation of AS-2 where CD and EF were associated was focused. The central position of CD and EF were first connected because of the two G-telemeric overhangs in G-quadruplex formation. After replacing with K^+-free buffer and introducing visible light irradiation, CD and EF

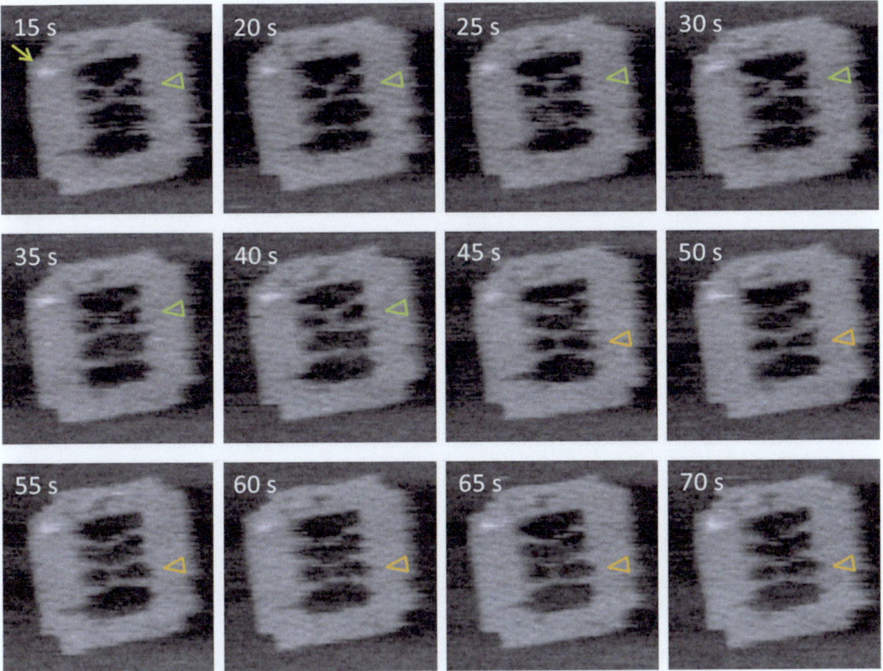

Fig. 3.6 The reversible AFM images of a single nanoframe's configuration changing from AS-2 to AS-1 under visible light irradiation in the observation buffer without K^+. AFM scanning rate was 0.2 frames per second. The lapsed time after starting photoirradiation was marked on each AFM image. The "kissing" between CD and EF was pointed by *orange triangles* while the "kissing" between AB and CD was pointed by *green triangles*. The hairpin marker was pointed by *yellow arrows*. Image size: 150 nm × 150 nm. [Yang, Y.; Endo, M.; Suzuki, Y.; Hidaka, K.; Sugiyma, H. *Chem. Comm.*, **2013**, 50, 4211–4213.] Reproduced by permission of The Royal Society of Chemistry

were observed to dissociate immediately and meanwhile AB and CD were asso-ciated in the middle of two dsDNAs because of the sequential hybridization of photoresponsive domains (40–45 s) in the reverse *cis-* to *trans-* isomerization of azobenzene molecules.

Besides the above immediate transition of two states, another type of changes from AS-2 to AS-1 (Fig. 3.7) were also observed, in which the CD and AB did not associate immediately after the deformation of G-quadruplex (95–110 s) without the help of potassium ion. The double-stranded CD vacillated between AS-1 and an unidentified state for around 15 s before stabilizing at AS-1 state finally (115 s). Interestingly, different mechanical movements of the interactions of three dsDNAs in single nanoframe were observed under different stimulating conditions. It was assumed that these differences might be mainly attributed to different association/dissociation manners of photoresponsive motifs and G-telomeric repeats. As a result, the sequential state transition between AS-2 and AS-1 were successfully visualized directly.

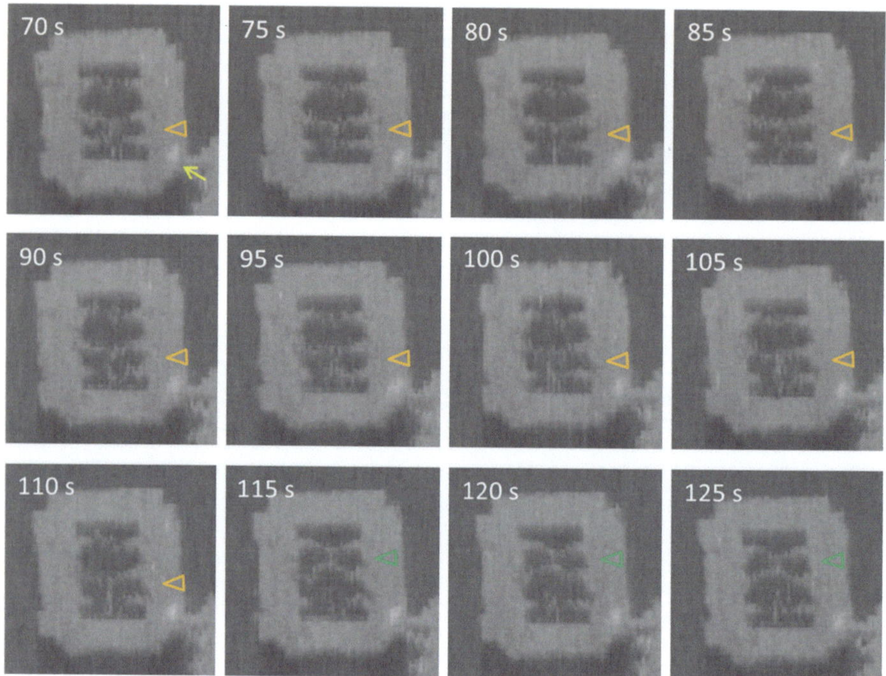

Fig. 3.7 Another obtained AFM images of a single nanoframe's configuration changing from AS-2 to AS-1 under visible light irradiation in the observation buffer without K$^+$. AFM scanning rate was 0.2 frames per second. The lapsed time after starting photoirradiation was marked on each AFM image. The association between CD and EF was pointed by *orange triangles* while the association between AB and CD was pointed by *green triangles*. Image size: 150 nm × 150 nm [Yang, Y.; Endo, M.; Suzuki, Y.; Hidaka, K.; Sugiyma, H. *Chem. Comm.*, **2013**, 50, 4211–4213.] Reproduced by permission of The Royal Society of Chemistry

3.4   Conclusion

In conclusion, a two-step cascade reaction of different single molecules was successfully implemented in a single DNA nanostructure by integrating two different external stimuli: photoregulation which can be controlled remotely and metal ion dependent switch (K$^+$). Furthermore, based on these two stimuli, the single logical conversions corresponding to conformational change were reversibly visualized in real time between finite-states by cooperating HS-AFM with DNA origami methodology. It displayed a potential approach for tracing single signal conversion in logic-controlled biosystems such as DNA computing. Although the response of the self-assembled molecular devices is basically slow compared with the electronic devices, these logical molecular systems can be applied for creating versatile devices that sense various external environments.

References

1. Seeman NC (2010) An. Rev Biochem 79:65–87
2. Endo M, Yang Y, Sugiyama H (2013) Biomater Sci 1:347–360
3. Qian L, Winfree E (2011) Science 332:1196–1201
4. Douglas SM, Bachelet I, Church GM (2012) Science 335:831–834
5. Zhu J, Zhang L, Li T, Dong S, Wang E (2013) Adv Mater 25:2440–2444
6. Liu H, Zhou Y, Yang Y, Wang W, Qu L, Chen C, Liu D, Zhang D, Zhu D (2008) J Phys Chem B 112:6893–6896
7. Prokup A, Hemphill J, Deiters A (2012) J Am Chem Soc 134:3810–3815
8. Okamoto A, Tanaka K, Saito I (2004) J Am Chem Soc 126:9458–9463
9. Li T, Ackermann D, Hall AM, Famulok M (2012) J Am Chem Soc 134:3508–3516
10. Chakraborty B, Jonoska N, Seeman NC (2012) Chem Sci 3:168–176
11. Costa Santini C, Bath J, Tyrrell AM, Turberfield A (2013) J Chem Comm 49:237–239
12. Wang Z-G, Elbaz J, Remacle F, Levine RD, Willner I (2010) Proc Nat Acad Sci 107: 21996–22001
13. Benenson Y, Gil B, Ben-Dor U, Adar R, Shapiro E (2004) Nature 429:423–429
14. Elbaz J, Wang Z-G, Orbach R, Willner I (2009) Nano Lett 9:4510–4514
15. Rajendran A, Endo M, Sugiyama H (2012) Angew Chem Int Ed 51:874–890
16. Sannohe Y, Endo M, Katsuda Y, Hidaka K, Sugiyama H (2010) J Am Chem Soc 132: 16311–16313
17. Endo M, Yang Y, Suzuki Y, Hidaka K, Sugiyama H (2012) Angew Chem Int Ed 51: 10518–10522
18. Rajendran A, Endo M, Hidaka K, Sugiyama H (2013) J Am Chem Soc 135:1117–1123
19. Ando T, Kodera N, Takai E, Maruyama D, Saito K, Toda A (2001) Proc Nat Acad Sci 98:12468–12472
20. Asanuma H, Liang X, Nishioka H, Matsunaga D, Liu M, Komiyama M (2007) Nat Protoc 2:203–212
21. Liang X, Mochizuki T, Asanuma H (2009) Small 5:1761–1768
22. Yang Y, Endo M, Hidaka K, Sugiyama H (2012) J Am Chem Soc 134:20645–20653
23. Darby RAJ, Sollogoub M, McKeen C, Brown L, Risitano A, Brown N, Barton C, Brown T, Fox KR (2002) Nucleic Acids Res 30:e39

Chapter 4
Multi-directional Assembly/Disassembly of Photocontrolled DNA Nanostructures in Programmed Patterns

4.1 Introduction

By specific complementary base pairing of nucleobases, DNA has been employed as versatile building blocks and contributes to the folded nanostructures in nanometer scale. DNA origami technology is a great progress for the structural DNA nanotechnology in the past decades [1]. DNA origami technique affords a convenient method for the construction of multi-dimensional nanoarchitectures in various shapes by using a long single-stranded DNA assembling with hundreds of helper oligonucleotides called staple strands. More importantly, the predesigned structures can be employed as nanometer-sized scaffolds for addressing with targeted molecules such as proteins [2–4], nanoparticles [5–8] and synthetic molecules at predefined positions for single molecule analysis [9, 10] or for further biological applications [11–13].

However, it is still a challenge for researchers to develop advanced programmable nanostructures under manual control. A few strategies have already been reported for the assembly of nanostructures in larger size or in multiple patterns. "Superorigami" method developed by Yan and his coworkers was applied for much larger spatial nanostructures by scaling up hundred-nanometer sized DNA origami tiles together using single-stranded DNA with bridge strands [14, 15]. Another representing method called "Jig-saw pieces" strategy was reported by our group, in which 1D and 2D programmed nanostructures were successfully assembled taking advantage of shape complementary of predesigned assembling units [16].

In this study, a distinctive method for organizing DNA origami units into designed regular or irregular programmed nanostructures is demonstrated by using a pair of pseudocomplementary photoresponsive oligonucleotides (Azo-ODN-1 and Azo-ODN-2) as assembling helpers to scale up into larger size. This pair of photocontrollable oligonucleotides modified with azobenzene molecules serve as not only the linkers to support the construction of larger nanostructures from origami

© Springer Japan 2015

Y. Yang, *Artificially Controllable Nanodevices Constructed by DNA Origami Technology*, Springer Theses, DOI 10.1007/978-4-431-55769-2_4

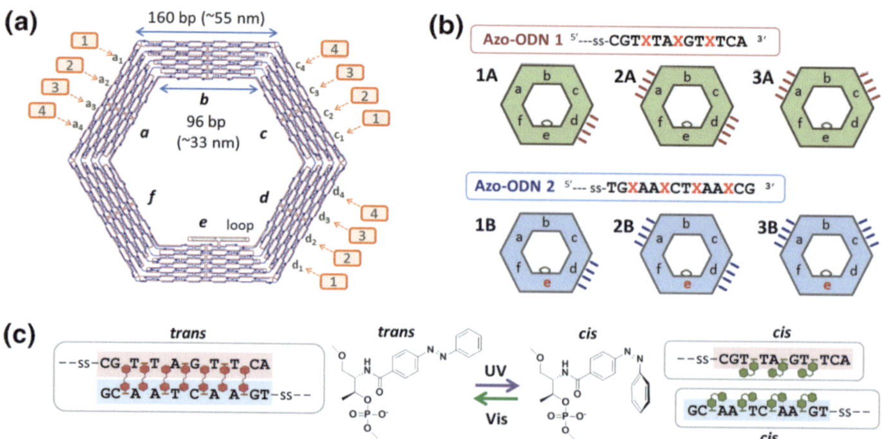

Fig. 4.1 Schematic drawings of photoresponsive DNA origami structure. **a** Design of hexagonal shaped unit in a size of ∼55 nm × ∼33 nm. Three edges (*a*-, *c*-, *d*-) was chosen for the modification with photoswitching motifs, where four staples strands of each edge were predetermined: a_1, a_2, a_3, a_4 and so forth. **b** Two strands: Azo-ODN 1 and Azo-ODN 2 employed as adhesive linkers were introduced the predetermined positions of hexagonal unit in three different kinds of arrangements: A-series in *green color* are modified with four Azo-ODN 1 strands and B-series in *blue color* are modified with four Azo-ODN 2 strands. To distinguish A and B series, a hairpin DNA markers were introduced to the e-domain of the B-series units which is marked by *red colored* "e". **c** Reversible photoswitching of the hybridization and dissociation of Azo-ODN 1 and Azo-ODN 2 under UV and visible light irradiation taking advantage of *trans-cis* photoisomerization of azobenzene moieties. Reprinted with the permission from Ref. [Yang, Y.; Endo, M.; Suzuki, Y.; Hidaka, K.; Sugiyama, H. J. Am. Chem. Soc., 2012, 51, 20645–20653.]. Copyright (2012) American Chemical Society

units but also as photoresponsive switches to regulate the assembly and disassembly of units reversibly in a wavelength-dependent manner [17–19]. The photosensitivity and photoregulated hybridization/dissociation of azobenzene-tethered strands in DNA nanostructures have already been confirmed in our previous studies by high-speed atomic force microscopy (AFM), showing the possibility of this pair of photoresponsive duplex can serve as adhesive switch at single molecule level.

As illustrated in Fig. 4.1, a hollow hexagonal-shaped DNA origami (∼55 nm × ∼33 nm, outer edge × inner edge) was designed as assembling unit, containing a small loop at the inner side of e-edge, which was employed as marker to distinguish the relative facing orientation (up or down) of unit in the further study. All six edges were all along helical axis to avoid π-stacking efficiently meanwhile afforded the possibility of multi-directional extension from outer edges. Three outer edges: a-, c- and d-, were selected and four strands of each edge were modified with switching strands separately (Fig. 4.1a). As shown in Fig. 4.1b, a pair of pseudocomplementary oligonucleotides: Azo-ODN 1 and Azo-ODN 2 were introduced into the predesigned positions with different arrangements, resulting six different patterns of hexagonal units carrying photoresponsive strands by disulfide bonds in two series (A-series and B-series). Each series is composed of three different modification types of hexagonal units: "d-edge" (**1A** and **1B**), "a-/d-edges" (**2A** and **2B**), and "a-/c-edges" (**3A** and **3B**) (Fig. 4.1b). A-series is modified by Azo-ODN 1 containing three azobenzene molecules while Azo-ODN 2 containing four azobenzene molecules is introduced into B-series. The pair of Azo-ODN 1 and Azo-ODN 2 can form the duplex in trans-form under visible light and dissociate in cis-form after UV-irradiation (Fig. 4.1c). The number and the addressing positions of Azo-ODN strands can be determined before assembling, which contributes for the following scaling up hexagonal units into different geometrical nanostructures. In this study, a series of photoregulated origami supernanostructures were intended to build up in regular and irregular arrangements.

4.2 Experimental Sections

4.2.1 *Modification of Staple Strands Along Hexagonal Edges with Photoresponsive Oligonucleotides*

The synthesis of staple strands with photoresponsive oligonucleotides was coupled by disulfide bond. The methods have already been demonstrated in Chap. 2.

4.2.2 *Assembly of Single Hexagonal Unit*

The hexagonal DNA origami was designed using caDNAno software. The unit was assembled in 20 μL of solution containing 10 nM M13mp18 single stranded DNA, 100 nM staple oligonucleotides (10 equiv), 20 mM Tris buffer (pH 7.6), 1 mM EDTA, and 10 mM $MgCl_2$. The mixture was annealed by reducing the temperature from 85 to 15 °C at a rate of -1.0 °C/min. The origami solution was purified using Sephacryl S-300 gel-filtration column after annealing finished. The hexagonal units bearing Azo-ODN strands were prepared by replacing the staples modified with Azo-ODN strands at the same position according to the design of different patterns of oligomers.

From the AFM images in Fig. 4.2, it can be seen that a hexagonal-shaped nanostructure is successfully assembled, in which "e-domain" of hexagonal monomer can be easily distinguished by the loop position clear imaged by AFM. In the larger-size AFM images, the hexagonal monomers were monodispersed without aggregation on mica surface.

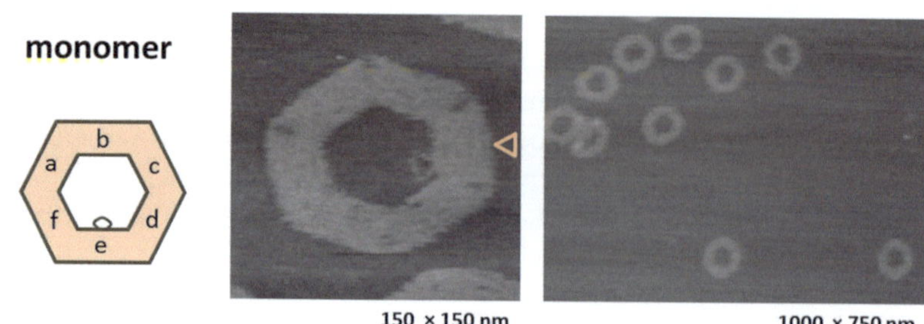

Fig. 4.2 Assembly of single unmodified hexagonal monomer and AFM images of assembled hexagonal monomer. "e-domain" pointed out by *orange triangle* in AFM image by the loop position. Reprinted with the permission from Ref. [Yang, Y.; Endo, M.; Suzuki, Y.; Hidaka, K.; Sugiyama, H. J. Am. Chem. Soc., 2012, 51, 20645–20653.]. Copyright (2012) American Chemical Society

4.2.3 Agarose Gel Electrophoresis Analysis

Hexagonal monomers were first prepared and used as makers to confirm the dimer dissociating into monomers. Photoirradiation to samples was carried out by light source of Xe-lamp (300 W, Ashahi-Spectra MAX-303) at 40 °C in a water bath. The light wavelength was switched by filter (UV: 350 nm, MX0350, Φ 25 mm; Visible: 450 nm, LX0450, Φ 25 mm, Asahi Spectra Co. Ltd.) After the photoirradiation, the samples were kept in the dark at room temperature. The samples were loaded to electrophoresis on a 0.6 % agarose gel containing 5 mM $MgCl_2$ in a 1 × TBE (Tris-borate-EDTA) buffer solution under 90 V at 4 °C. The gels were then imaged using ethidium bromide as staining dye by Wealtec Dolphin-View 2 (KURABO Industries, Ltd.). The intensities of bands were quantified by Image J software (NIH).

4.2.4 Fluorescent Analysis of Disassemble/Assemble of Hexagonal Dimers

Fluorescence spectra were measured on JASCO FP-8300 with thermal-controlling mode. Photoirradiation to the samples was performed in a microcell placed in the cell holder of the spectrometer by keeping the temperature at 40 °C. The spectrum was directly measured after the photoirradiation. Excitation wavelength for FAM was 490 nm and fluorescence emission was recorded at 516 nm.

4.3 Results and Discussions

By using different combinations of hexagonal units from A-series and B-series as aggregation units, various patterns of multimer such as in linear shape, curved shape and ring shape, are planned to assemble with different facing orientations.

4.3.1 Preparation and Characterization of Hexagonal Dimer

The hexagonal dimer was first tried to construct by employing hexagonal units containing Azo-ODN strands: 1A containing four Azo-ODN 1 strands at d-edge and 1B containing another four Azo-ODN 2 strands also located at d-edge. Without any irradiation treatment, the azobenzene molecules of Azo-ODN 1 and Azo-ODN 2 remain in trans-form, which will associate with each other and form the duplex. As a result, the hexagonal dimer will be formed with linkage of four Azo-ODN duplexes. The mechanism of dimer formation was illustrated in Fig. 4.3.

After the preparation two selected monomers separately, they were coupled into dimer by mixing equivalent 1A (5 nM) and 1B (5 nM) in same assembling buffer (20 mM Tris buffer (pH 7.6), 1 mM EDTA, and 10 mM MgCl$_2$) by annealing from

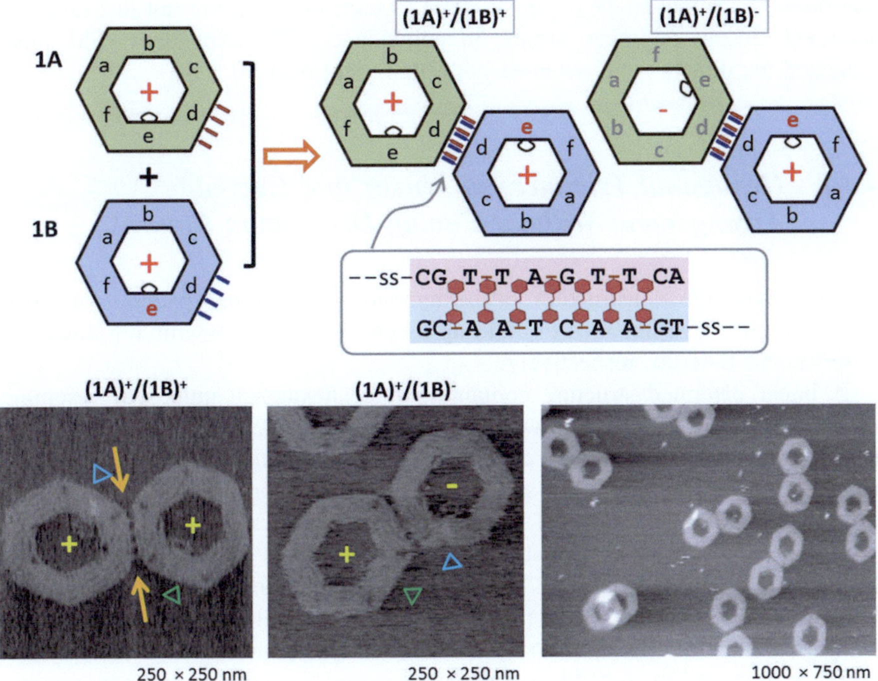

Fig. 4.3 Two hexagonal monomers from A- and B- series, separately (1A and 1B) were assembled into hexagonal dimer, in which each monomer carrying four Azo-ODN strands along one edge. Hairpin markers were introduced into 1B unit to distinguish two different monomers, and the facing orientations of two monomers in a dimer could be distinguished by comparing the relative position of loops at e-domain of 1A and 1B (pointed by *blue* and *green triangles*). The four hybridized Azo-ODN duplexes connecting two monomers were directly imaged by AFM, which are pointed out by two orange arrows in AFM image. Reprinted with the permission from Ref. [Yang, Y.; Endo, M.; Suzuki, Y.; Hidaka, K.; Sugiyama, H. J. Am. Chem. Soc., 2012, 51, 20645–20653.]. Copyright (2012) American Chemical Society

50 to 15 °C in a rate of −0.1 °C/min. The assembled structure was then directly confirmed by AFM without further purification. The results were shown in Fig. 4.3. The hexagonal monomers: 1A and 1B are both geometrically symmetrical because four Azo-ODN strands were tethered along one edge, resulting that two types of hexagonal dimers will be formed owing to that the four parallel duplex linkages are same inducing the dimer are lack of relative facing up/down control. Two types of dimer with different facing orientation of were obtained. One type is of same facing orientation ("+/+", Fig. 4.3) and another type is of opposite facing orientation ("+/−", Fig. 4.3). The relative facing orientation should be distinguished by comparing relative positions of the loops at the inner side of e-domain (marked on AFM images by blue and green triangles in Fig. 4.3). In addition, hairpin markers were introduced to the e-domain for the preparation of B-series monomer to differentiate assembling units of A-series and B-series. In the AFM images of dimers, four Azo-ODN duplexes as linkers between two monomers were also clearly imaged in high resolution (pointed by orange arrows in AFM image), representing that these hexagonal-shaped nanostructures were in dimer form not just randomly absorbed closely on mica surface in coincidence. The expanded AFM image indicated the dimers were obtained with high yield over 90 %.

4.3.2 Hexagonal Trimmers in Linear and Curved Arrangement Without Facing Orientation Control

Learned from the construction of hexagonal dimers, other patterns of longer nanostructures were intended to build up by choosing other assembling units from A-series and B-series, respectively.

A linear patterned structure containing three hexagonal units was assembled utilizing 2A-unit and 1B-unit. Two edges (a- and d-) in 2A were both extended with four Azo-ODN 1 strands along each edge, which can integrate with two 1B units containing four Azo-ODN 2 strands at d-edge to form a trimmer in linear arrangement, shown in Fig. 4.4. Same as the symmetrical property of hexagonal dimer, there should be three patterns of linear trimmers without facing up/down control: "(1B)−/(2A)+/(1B)−", "(1B)+/(2A)−/(1B)−", and "(1B)+/(2A)+/(1B)+". Experimentally, the trimmers were prepared by mixing already assembled 2A and 1B in a ratio of 1: 2. The solution was then annealed from 50 to 15 °C in a rate of −0.05 °C/min and then AFM imaging was carried out for the characterization of assembled nanostructures. As expected, three patterns of linear trimmer with three kinds of relative facing orientations were obtained in the same annealing solution. As shown in Fig. 4.1a, two 2B monomers in same facing orientation were located at the two sides of 1A which is in opposite direction of 2B. The assembling unit, 1A and 2B, can be identified by the hairpin marker. In Fig. 4.4b, a 1B and a 2A were assembled in same orientation while another 1B were in opposite direction. In these two types of trimmer, one 1B unit and 2A unit were in mirror symmetry. The facing

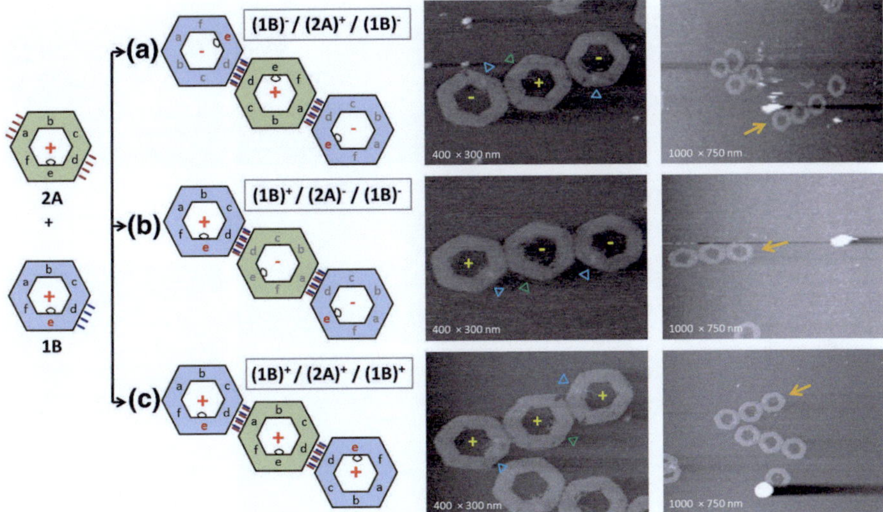

Fig. 4.4 Schematic illustrations and AFM images of hexagonal trimmers in the linear arrangement using 2A and 1B as self-assembly substrates. Three types of trimmers without facing up/down control were obtained and confirmed by AFM: **a** "(1B)−/(2A)+/(1B)−", **b** "(1B)+/(2A)−/(1B)−", **c** "(1B)+/(2A)+/(1B)+". The relative facing orientations of three types of linear trimmers were distinguished by the three individual positions of loops at the e-domain. Reprinted with the permission from Ref. [Yang, Y.; Endo, M.; Suzuki, Y.; Hidaka, K.; Sugiyama, H. J. Am. Chem. Soc., 2012, 51, 20645–20653.]. Copyright (2012) American Chemical Society

orientations of three monomers of third type of trimmer (Fig. 4.4c) were found out to be all the same. Here the relative facing orientation of each monomer was determined by the individual loop position in e-domain of assembling substrate. The linear patterned trimmers were successfully obtained in three kinds of relative facing orientations by employing 2A and 1B as assembling substrate.

On the other hand, another kind of trimmer in curved pattern were tried to construct based on similar rules of linear trimmer by replacing 2A unit to 3A unit. 3A unit was extended by Azo-ODN 1 strands along a- and c-edge which were in an angle of 120°, affording the possibility of non-linear accumulation of complementary hexagonal units such as 1B. Under the same annealing conditions for linear trimmers, the construction of curved trimmers was implemented by mixing 3A unit with 1B in a ratio of 1: 2. The predicted structure and AFM images structures were shown in Fig. 4.5. As a result, three types of curve-patterned trimmers were obtained with three kinds of relative facing orientations same as linear trimmers. In same annealing solution, different types of trimmers in one pattern but in different relative facing up/down directions were successfully constructed by employing predesigned assembling units from both A-series and B-series, owing to the symmetry of hexagonal unit containing four Azo-ODN strands.

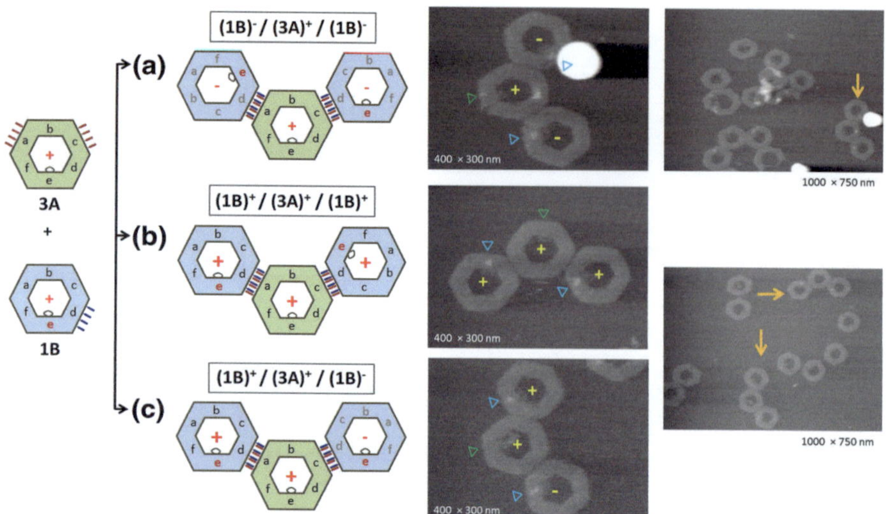

Fig. 4.5 Schematic illustrations and AFM images of hexagonal trimmers in the curved arrangement using 2A and 1B as self-assembly substrates. Three types of trimmers without facing up/down control were obtained and confirmed by AFM: **a** "(1B)−/(3A)+/(1B)−", **b** "(1B)+/(3A)+/ (1B)+", **c** "(1B)+/(3A)+/(1B)−". The relative facing orientations of three types of curved trimmers were distinguished by the three individual positions of loops at the e-domain. Reprinted with the permission from Ref. [Yang, Y.; Endo, M.; Suzuki, Y.; Hidaka, K.; Sugiyama, H. J. Am. Chem. Soc., 2012, 51, 20645–20653.]. Copyright (2012) American Chemical Society

Here agarose gel electrophoresis analysis was used to quantify the yield of assembled nanostructures. As shown in Fig. 4.6, two slowest bands were imaged indicating the formation of linear trimmers and curved trimmers in high yields.

4.3.3   Hexagonal Oligomers Without Facing Orientation Control

Using hexagonal units: 2A and 2B containing Azo-ODN strands at two edges of one unit the hexagonal oligomer were tried to assemble. As illustrated in Fig. 4.7, the linear-patterned nanostructure can be formed. After mixing the two monomers 2A and 2B in the same equivalent, the annealing of self-assembling was first started non-linearly from 50 to 35 °C by decreasing the temperature by 6 °C in a rate of −0.01 °C/min and increasing it by 3 °C in a rate of +0.5 °C/min and then gradually decreased from 35 to 15 °C linearly in a rate of −0.01 °C/min. The annealed solution was then characterized imaged by AFM. Four linear hexagonal oligomers in different lengths were shown in Fig. 4.7, in which the 2A and 2B connected with each other alternatively by the formation of Azo-ODN duplexes tethered along two

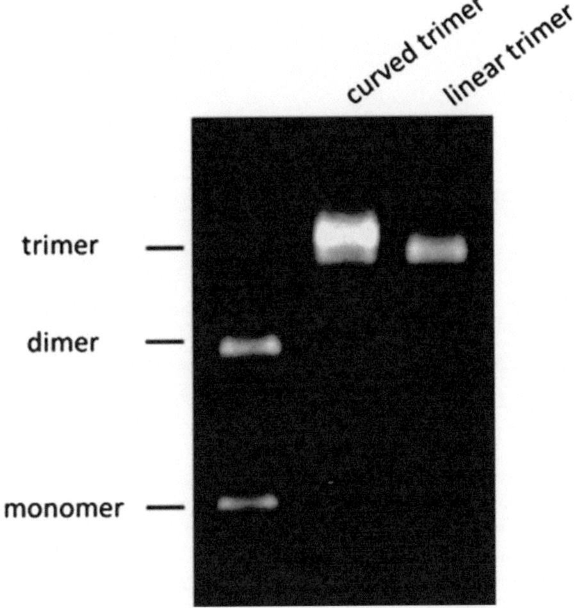

Fig. 4.6 Agarose gel electrophoresis analysis (0.5 % agarose containing 5 mM $MgCl_2$, 1 × TBE buffer) of the linear trimmer and curved trimmers. First lane is the mixture of dimer and monomer employed as the marker. The yields of linear trimmers (70 %) and curved trimmers (80 %) were then analyzed and quantified by Image J. Reprinted with the permission from Ref. [Yang, Y.; Endo, M.; Suzuki, Y.; Hidaka, K.; Sugiyama, H. J. Am. Chem. Soc., 2012, 51, 20645–20653.]. Copyright (2012) American Chemical Society

edges. The longest linear structures on mica surface were composed of 7 hexagonal monomers.

Besides, the long curve-shaped oligomers were also assembled by using 3A and 3B as accumulation units (Fig. 4.8). Because of the same symmetry of 3A and 3B, the obtained structures were in various irregular zig-zag patterns, shown in Fig. 4.8. Most of the curved structures contained four or five assembling units.

Each hexagonal unit had two assembling orientations, resulting that the relative facing orientations of both the linear oligomers and curved oligomers were in random arrangements. In AFM images, it can be see that the loops of hexagonal unit were found in random.

The agarose gel electrophoresis analysis was carried out to evaluate the assembled hexagonal oligomers in linear and curved patterns. As shown in Fig. 4.9b, two samples of linear oligomer and curved oligomer were observed as two smeared bands (lane 1 and lane 2) by using monomer, dimer and trimmer as marker, indicating that the hexagonal oligomers were formed in different numbers of units. Only short oligomers were imaged by AFM probably because of the potential fracture of the longer oligomers during the loading onto the mica surface. The indirect gel analysis and direct AFM imaging showed different results of

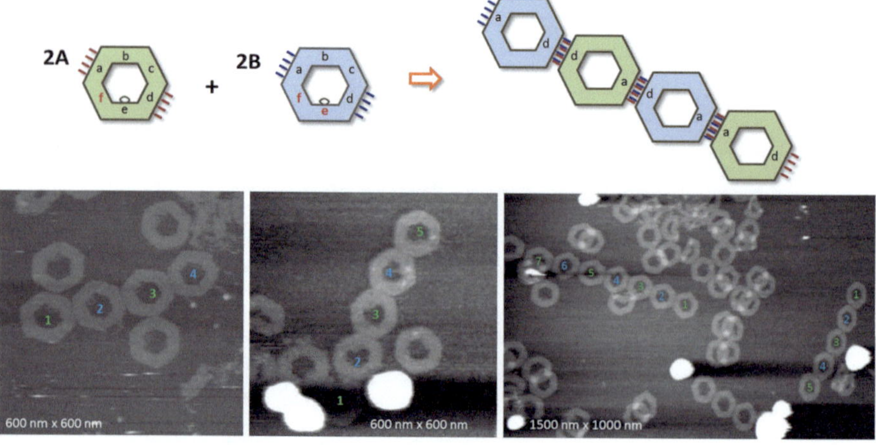

Fig. 4.7 Schematic illustrations and AFM images of hexagonal oligomers in linear arrangement by using 2A and 2B as assembling unit. A hairpin marker was introduced to the f-domain of 2A and two hairpin markers were introduced to e-/f- domains of 2B. Four linear hexagonal oligomers in different lengths were imaged by AFM. 2A and 2B were connected alternatively by Azo-ODN duplexes. Reprinted with the permission from Ref. [Yang, Y.; Endo, M.; Suzuki, Y.; Hidaka, K.; Sugiyama, H. J. Am. Chem. Soc., 2012, 51, 20645–20653.]. Copyright (2012) American Chemical Society

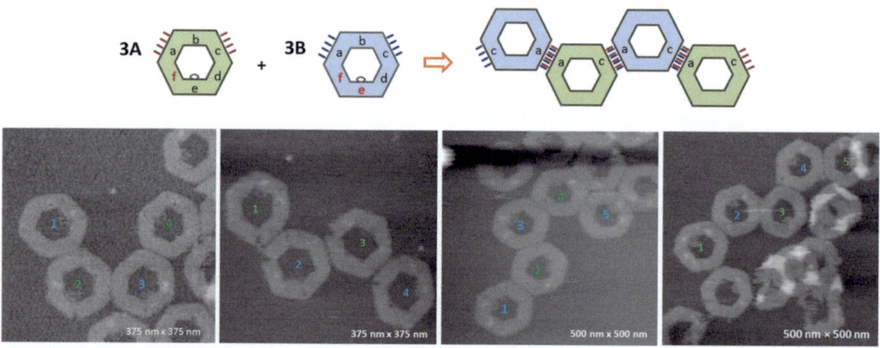

Fig. 4.8 Schematic illustrations and AFM images of hexagonal oligomers in curved arrangement by using 3A and 3B as assembling unit. A hairpin marker was introduced to the f-domain of 3A and two hairpin markers were introduced to e-/f- domains of 3B. Four different hexagonal oligomers in irregular zig-zag pattern were obtained by AFM. 3A and 3B were connected alternatively by Azo-ODN duplexes. Reprinted with the permission from Ref. [Yang, Y.; Endo, M.; Suzuki, Y.; Hidaka, K.; Sugiyama, H. J. Am. Chem. Soc., 2012, 51, 20645–20653.]. Copyright (2012) American Chemical Society

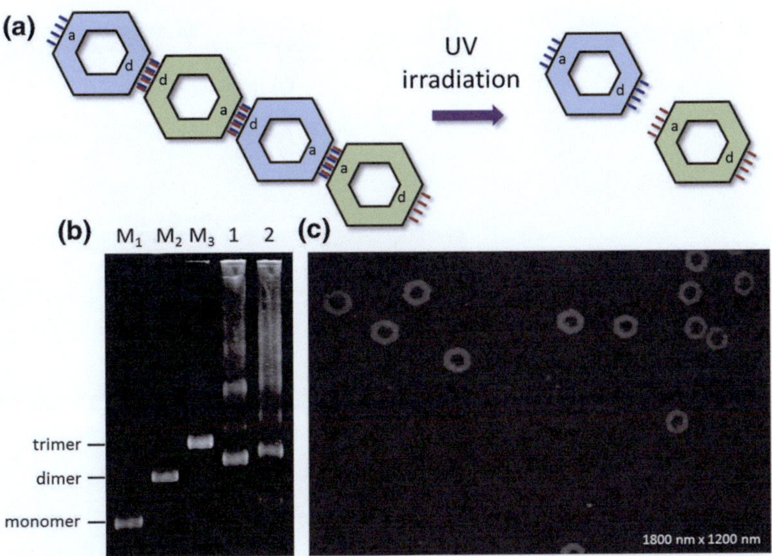

Fig. 4.9 (a) Schematic illustration of dissembling of linear oligomers into single monomers under UV irradiation. (b) Agarose gel (0.4 %) image of assemblies from 2A and 2B (lane 1) and 3A and 3B (lane 2). Lanes M1, M2, and M3: unmodified monomer, dimer (1A−1B), and linear trimmer (1B−2A−1B), respectively were employed as markers. (c) AFM image of linear 2A−2B oligomers after UV irradiation treatment. UV irradiation was carried out for 5 min at 35 °C. Reprinted with the permission from Ref. [Yang, Y.; Endo, M.; Suzuki, Y.; Hidaka, K.; Sugiyama, H. J. Am. Chem. Soc., 2012, 51, 20645–20653.]. Copyright (2012) American Chemical Society

hexagonal oligomers, presenting that both of two analyzing methods were necessary for the characterization of the nanostructures.

Figure 4.9a demonstrated the photoregulated dissemble of linear oligomer into single monomers. The Azo-ODN duplexes are photosensitive, which can dissociate after the UV-induced photoisomerization of azobenzene molecules. AFM imaging confirmed the UV-treated linear oligomers (Fig. 4.9c), most of the hexagonal monomers were imaged on mica surface, indicating that the assembled oligomeric structures can be easily dissembled into monomers efficiently ($\sim$90 % of conversion).

4.3.4   Hexagonal Oligomers in Linear and Curved Arrangements with Facing Orientation Control

Next it was attempted to construct the predesigned nanostructures with facing control of each assembling unit by adjusting the numbers of Azo-ODN linkers in hexagonal unit. Finally three Azo-ODN linkers were placed asymmetrically along

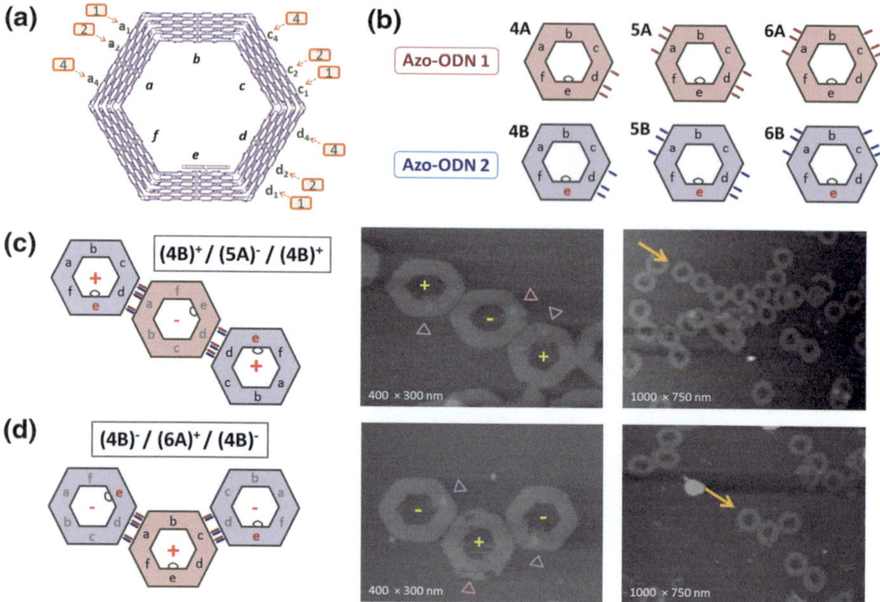

Fig. 4.10 Schematic drawings and AFM images of hexagonal trimmers in both linear and curved arrangement with facing up/down control. **a** Design of asymmetrical hexagonal unit by introducing three Azo-ODN strands at the position of 1, 2, and 4 along every edge of hexagonal unit. **b** Six different patterns of hexagonal monomers in two series (A- and B-) were further designed. For identification of hexagonal monomers, hairpin DNA markers were introduced to the e-domain of the B-series monomers (marked by *red colored* "e"). **c** Linear trimmers with facing up/down control in the arrangement of "(4B)+/(5A)−/(4B)+" were assembled from 4B-unit and 5A-unit and imaged by AFM. **d** Curved trimmers with facing control in the arrangement of "(4B)/(6A)+/(4B) −" were assembled from 4B-unit and 6A-unit and imaged by AFM. The relative facing orientation of trimmers could be distinguished by the three relative positions of the loops (pointed out by *pink* and *purple triangles*) at the e-domain of each unit. Reprinted with the permission from Ref. [Yang, Y.; Endo, M.; Suzuki, Y.; Hidaka, K.; Sugiyama, H. J. Am. Chem. Soc., 2012, 51, 20645–20653.]. Copyright (2012) American Chemical Society

each edge at the position 1, 2 and 4 (Fig. 4.10a). Based on this design, the other six assembly units was determined and divided into A-series and B-series, listed in Fig. 4.10b. Hexagonal unit 4B and 5A were employed for assembling of linear trimmer, and the other trimmer in curved pattern was constructed from 4B and 6A (Fig. 4.10c). Because of the asymmetry of selected hexagonal units, the facing orientation can be controlled precisely in the full-matching trimmer formation. As a result, the linear trimmer is in the geometrical formation of "(4B)+/(5A)−/(4B)+" while the curved trimmer is "(4B)+/(6A)−/(4B)+". The experimental preparation of the directional controlled trimmer was the same as the above multidirectional trimmers. The AFM imaging confirmed that both trimmers were formed correctly and no other structures in different facing orientations were observed (Fig. 4.10c, d).

The relative facing directions of each single unit can be strictly controlled based on our asymmetrical design. Moreover, by removing one pair of Azo-ODN duplex does not affect the formation of expected structure.

4.3.5 Large Hexagonal Oligomers with Restricted Facing Orientation Control

Besides the success construction of directionally restricted trimmers in both curved and linear patterns, the programmed linear oligomers was attempted to assemble by using 5A and 5B. 5A and 5B were also modified with Azo-ODN strands in asymmetrical placement. In Fig. 4.11, to make sure the two complementary units assembled completely, 5A and 5B need to be arranged in an alternative up-down arrangement. Under same nonlinear followed by linear annealing conditions as the above oligomers connected by four Azo-ODN duplexes, the sample was further confirmed by AFM. As a result, the oligomers in linear patterns were obtained. The longest structure was composed of seven units. The facing orientation of each assembling unit can be distinguished by the loop position of e-domain. However, there were two possible connection modes. One is a/a or d/d connection, another is a/d connection (illustrated in the orange dashed box in Fig. 4.11). But the alternative facing orientation $(+/-)$ will not be affected.

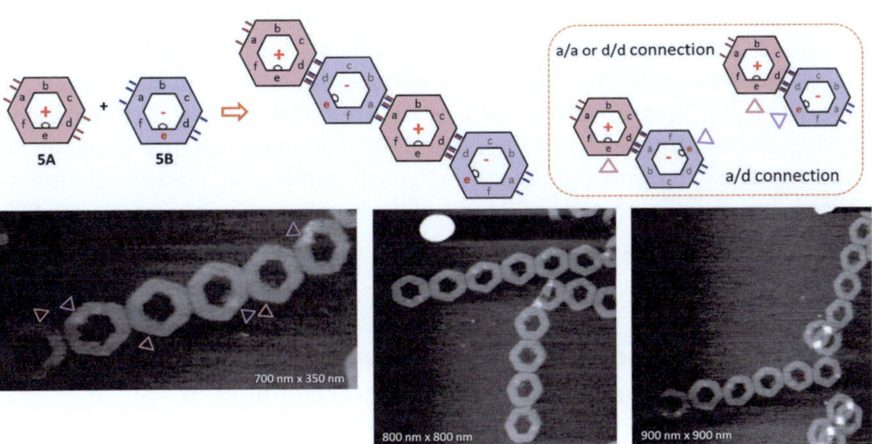

Fig. 4.11 Design and AFM images of assembled linear oligomers in an arrangement of alternative 5A-5B-5A-5B with controlled facing orientation $(+/-)$. *Dashed orange box*: two kinds of assembling modes: "a/a or d/d" and "a/d" connection. The relative facing orientation of each unit can be distinguished by the loop position located at the e-domain. Reprinted with the permission from Ref. [Yang, Y.; Endo, M.; Suzuki, Y.; Hidaka, K.; Sugiyama, H. J. Am. Chem. Soc., 2012, 51, 20645–20653.]. Copyright (2012) American Chemical Society

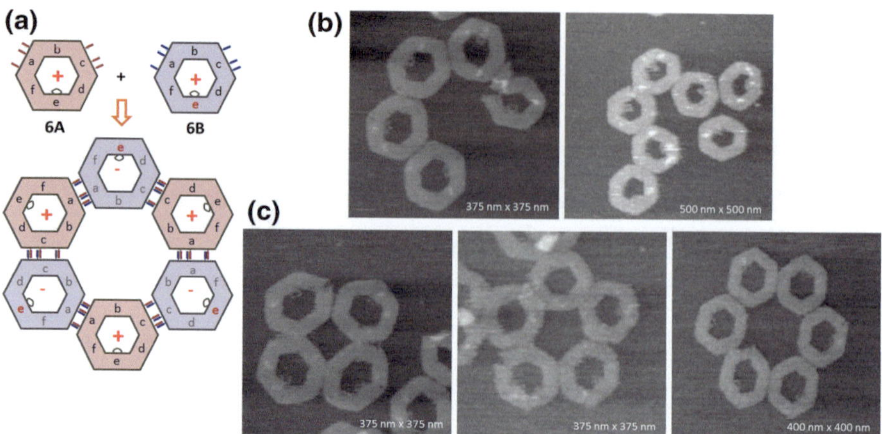

Fig. 4.12 Schematic drawings and AFM images of hexagonal oligomers in various open curved and closed ring arrangements with facing up/down control using the hexagonal units 6A and 6B. Reprinted with the permission from Ref. [Yang, Y.; Endo, M.; Suzuki, Y.; Hidaka, K.; Sugiyama, H. J. Am. Chem. Soc., 2012, 51, 20645–20653.]. Copyright (2012) American Chemical Society

The curved oligomers with restricted directional control were also prepared by employing 6A and 6B (Fig. 4.12a). Based on the asymmetrical placement of connecting motifs and the geometrical property of hexagonal unit, a ring-shaped structure containing six units (three 6A and 6B) with alternative facing control was anticipated. After the AFM imaging, a various patterns of oligomers were obtained in the same annealed solution. Figure 4.12b listed two kinds of curved open structures containing four units and five units separately. Figure 4.12c showed three ring-shaped structures. As expected, a hexagonal ring was finally successfully constructed. But because of the elasticity of Azo-ODN duplex ($\sim$10 nm length), the ring-shaped structures were composed of four units and five units. In this section, a various patterns of hexagonal oligomers with facing up/down control was realized by using asymmetrical assembling units containing three Azo-ODN strands.

4.3.6   Evaluation of Reversible Assemble and Disassembly of Hexagonal Dimers by Gel Electrophoresis

The Azo-ODN 1 and Azo-ODN 2 were not only pseudocomplementary to each other but can be reversibly regulate to hybridize and dissociate by switching the photoirradiation between UV (350 nm) and visible light (450 nm). So it is prospected to regulate the assembly and disassembly of the accumulated oligomers. The disassembly of linear oligomers has already been verified. In this part, it is mainly

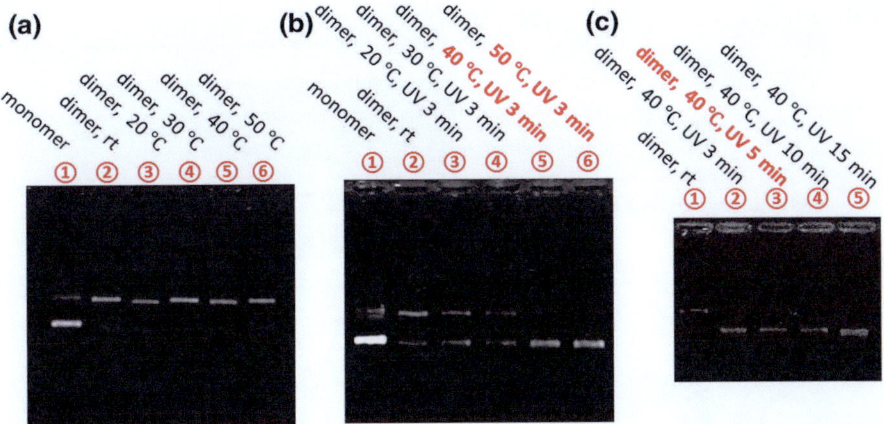

Fig. 4.13 Effect of temperature and UV irradiation time on the disassembling of hexagonal dimers into monomers analyzed by agarose gel electrophoresis. **a** Effect of temperatures on the structural stability of hexagonal dimers; lane 1 was used as marker for the hexagonal monomers. **b** Effect of temperatures on the disassembling of dimers under the same UV irradiation time (5 min); monomers were loaded into lane 1 as marker; lane 2 was employed as control. **c** Effect of different UV irradiation time to dissociate the dimer under 40 °C by water bath. Agarose gel electrophoresis analysis conditions: 0.6 % agarose containing 5 mM $MgCl_2$, 1 × TBE buffer, 90 V, 4 °C. Reprinted with the permission from Ref. [Yang, Y.; Endo, M.; Suzuki, Y.; Hidaka, K.; Sugiyama, H. J. Am. Chem. Soc., 2012, 51, 20645–20653.]. Copyright (2012) American Chemical Society

focused on the investigation of reversible photoregulated performances of assembled nanostructures. The hexagonal dimer was employed as model system for examining the reversible self-assemble performances (Fig. 4.14a). Firstly, a static analysis for dimers was carried out by using agarose gel electrophoresis. To find out optical experimental conditions such as temperature and photoirradiation time to regulate the association and dissociation of dimers, we evaluated photoresponsive efficiency of disassembly by regulating the different irradiation time under different temperatures. The results were summarized in Fig. 4.13. Figure 4.13a demonstrated that there was almost no breakage of the dimers even the temperature was as high as 50 °C. After introducing the UV irradiation (3 min), the dimer can be totally disassembled into monomers at relative high temperature (≥40 °C). Increasing temperature and extending irradiation time could both contribute to the disassembly of dimers. Figure 4.13c revealed the optical conditions of irradiation time as well as temperature: 40 °C, 5 min.

A continuous photoirradiation was introduced to dimer by switching UV and visible light alternatively by band pass filters (each process for 5 min). And the light-treated sample was analyzed by agarose gel electrophoresis. As can be seen in Fig. 4.14b, the hexagonal dimer can be regulated reversibly in the formation of dimer (vis) and monomer (UV) with relatively high yield (from lane 1 to lane 7).

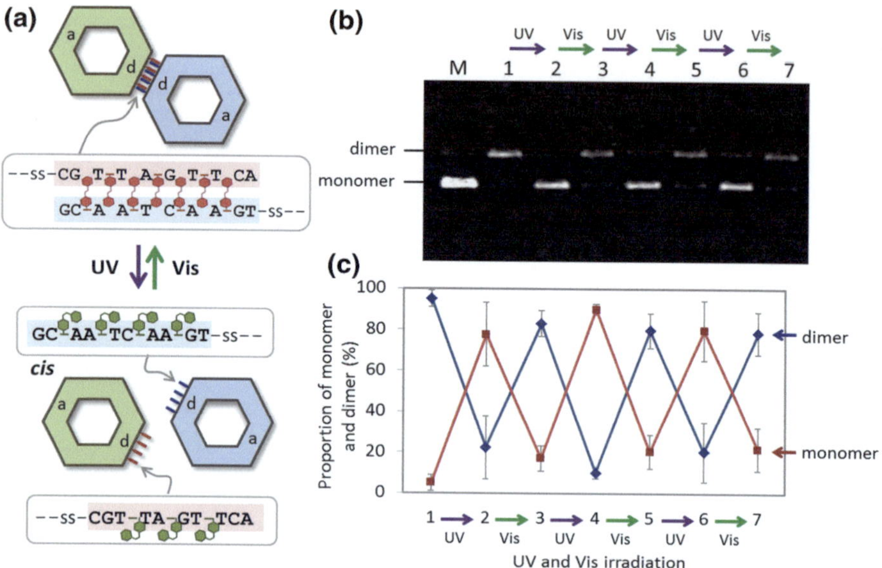

Fig. 4.14 Reversible photoregulation for the formation and disassembly of hexagonal dimer by switching photoirradiation between UV and visible light range. **a** Schematic drawings of reversible photoregulation of dimers. **b** Agarose gel (0.6 %) electrophoresis analysis of the photoregulation for the formation and dissociation between dimer and monomer by UV and visible light (Vis) irradiation. **c** Proportion of dimer and monomer under different irradiation conditions by quantification of the band intensity in (**b**), which were evaluated by Image J. Lane M represented the monomer without modification employed as a marker. Error bar indicated three repeated experiments. Reprinted with the permission from Ref. [Yang, Y.; Endo, M.; Suzuki, Y.; Hidaka, K.; Sugiyama, H. J. Am. Chem. Soc., 2012, 51, 20645–20653.]. Copyright (2012) American Chemical Society

The UV induced disassembly of dimer into monomer was caused by the dissociation of Azo-ODN duplexes with the photoisomerization of azobenzene molecules. The UV induced monomer could reassemble after switching to visible light. The gel images were further analyzed with the help of Image J to evaluate the proportion of dimer and monomer during the photoirradiation, shown in Fig. 4.14c. It can be seen that there were still 80 % of dimer formed even after three cycles of irradiation. And after seven cycles of UV/Vis photoirradiation, we can see that the dimer can still be formed with relative high yield. These results clearly demonstrated that Azo-ODN duplex could efficiently control the formation of hexagonal dimer by remote photoirradiation. Furthermore, it can be deduced that other programmed patterns of hexagonal oligomers should be also performed in similar fashion under the different photo irradiation conditions.

4.3.7 Dynamic Analysis of Assemble and Disassembly of Photocontrolled Hexagonal Dimers

Besides the static analysis, the fluorescence measurement was employed to real-time analyze the formation and deformation of dimers during the photoirradiation. The design of fluorescently labeled monomers was illustrated in Fig. 4.15a. The c/d corner of 1A was extended at 3' terminus by 14 nucleotide (nt) overhang containing 8 nt T spacer and 6 nt stretch tailed with BHQ-1 as quenching group. The other complementary unit was modified at 5' terminus of d/e corner by the 8 nt T spacer and the complementary 6 nt tethered with FAM. The fluorescence of FAM will be quenched in the dimer state under same facing directions by the irradiation with visible light (fluorescence OFF). And the fluorescence will be recovered after disassembling into monomer state by UV light. To increase the quenching efficiency in the dimer state, here three Azo-ODNs in asymmetrical arrangement (shown in Fig. 4.15a) were introduced into each kind of unit preferring the formation of dimers in relative same facing direction (" +/ +").The fluorescence spectra changes were shown in Fig. 4.15b under one cycle of photoirradiation (control-UV–Vis). The real-time fluorescence (λ_{ex}: 490 nm; λ_{em}: 516 nm) was taken

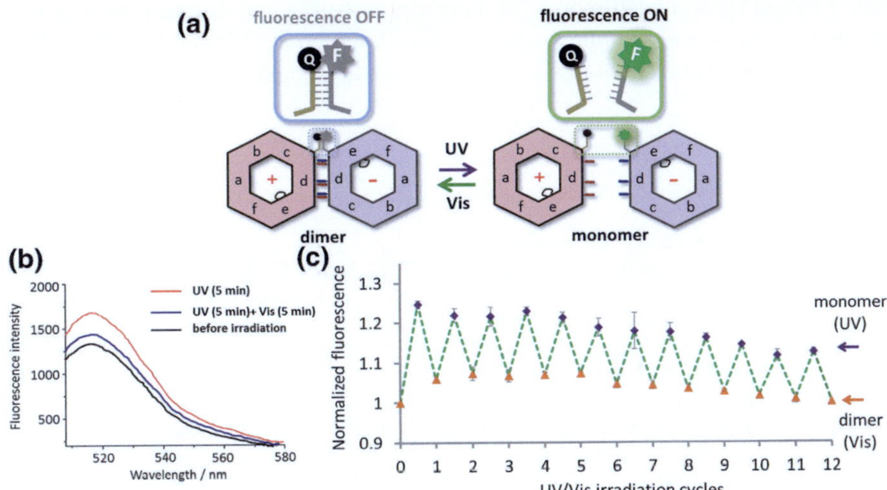

Fig. 4.15 Dynamic analysis of dimer formation and dissociation in solution under different irradiation conditions in real-time. **a** Schematic of two monomer units carrying FAM (5'-end) and BHQ-1 (3'-end) separately. Fluorescence OFF: in dimer formation under visible light; Fluorescence ON: dissociated into two monomers under UV irradiation. **b** Reversible normalized fluorescence intensity changes in dimer and monomer states. The numbers of horizontal axis referred to UV/vis irradiation cycles. **c** One cycle of fluorescence spectral change of the formation and dissociation of hexagonal dimer. Conditions: excitation, 490 nm; emission, 516 nm; 40 °C. Reprinted with the permission from Ref. [Yang, Y.; Endo, M.; Suzuki, Y.; Hidaka, K.; Sugiyama, H. J. Am. Chem. Soc., 2012, 51, 20645–20653.]. Copyright (2012) American Chemical Society

and the normalized fluorescence was listed in Fig. 4.15c. By measuring and comparing the fluorescence intensity after the UV and visible light irradiation, we can see that even over 10 rounds of irradiation cycle, the dimers assembling and disassembling still work well. About all, it is noteworthy that such reversible assembly and disassembly was not only the re-constituting repeatedly of two pieces, but also more significantly this photo-regulating process was a recombination of each specific unit in homology similar to the card shuffling.

4.4   Conclusion

A ~50 nm-sized hexagonal DNA origami was designed and constructed, which further employed as assembling unit to form larger-sized nanostructure in various 2D patterns by modifying the origami edges with photoresponsive oligonucleotides (Azo-ODN 1 and Azo-ODN 2). Furthermore, the aggregated multiple nanostructures can be regulated to disassemble and assemble by different wavelength of photoirradiation based on the hybridization and dissociation of the photoresponsive duplexes. Interestingly, the programmed facing orientations of each assembling unit can be manually controlled by regulating the number and the position of photoresponsive oligonucleotides. These newly designed photo-responsive DNA nanostructures can be regulated repeatedly between assembly and disassembly states by photo irradiation in a shuffling fashion, which shows great potential for the practical applications in nanotechnology, such as nanomedicine and nanoelectronics.

References

1. Rothemund PWK (2006) Nature 440:297–302
2. Endo M, Katsuda Y, Hidaka K, Sugiyama H (2010) Angew Chem Int Ed 49:9412–9416
3. Nakata E, Liew FF, Uwatoko C, Kiyonaka S, Mori Y, Katsuda Y, Endo M, Sugiyama H, Morii T (2012) Angew Chem Int Ed 51:2421–2424
4. Saccà B, Meyer R, Erkelenz M, Kiko K, Arndt A, Schroeder H, Rabe KS, Niemeyer CM (2010) Angew Chem Int Ed 49:9378–9383
5. Zhao Z, Jacovetty EL, Liu Y, Yan H (2011) Angew Chem Int Ed 50:2041–2044
6. Endo M, Yang Y, Emura T, Hidaka K, Sugiyama H (2011) Chem Commun 47:10743–10745
7. Kuzyk A, Schreiber R, Zhang H, Govorov AO, Liedl T, Liu N (2014) Nat Mater 13:862–866
8. Kuzyk A, Schreiber R, Fan Z, Pardatscher G, Roller E-M, Hogele A, Simmel FC, Govorov AO, Liedl T (2012) Nature 483:311–314
9. Yoshidome T, Endo M, Kashiwazaki G, Hidaka K, Bando T, Sugiyama H (2012) J Am Chem Soc 134:4654–4660
10. Yun JM, Kim KN, Kim JY, Shin DO, Lee WJ, Lee SH, Lieberman M, Kim SO (2012) Angew Chem Int Ed 51:912–915
11. Li J, Pei H, Zhu B, Liang L, Wei M, He Y, Chen N, Li D, Huang Q, Fan C (2011) ACS Nano 5:8783–8789
12. Douglas SM, Bachelet I, Church GM (2012) Science 335:831–834
13. Jiang Q, Song C, Nangreave J, Liu X, Lin L, Qiu D, Wang Z-G, Zou G, Liang X, Yan H, Ding B (2012) J Am Chem Soc 134:13396–13403

14. Zhao Z, Liu Y, Yan H (2011) Nano Lett 11:2997–3002
15. Zhao Z, Yan H, Liu Y (2010) Angew Chem Int Ed 49:1414–1417
16. Rajendran A, Endo M, Katsuda Y, Hidaka K, Sugiyama H (2010) ACS Nano 5:665–671
17. Asanuma H, Liang X, Nishioka H, Matsunaga D, Liu M, Komiyama M (2007) Nat Protoc 2:203–212
18. Liang X, Mochizuki T, Asanuma H (2009) Small 5:1761–1768
19. Endo M, Yang Y, Suzuki Y, Hidaka K, Sugiyama H (2012) Angew Chem Int Ed 51:10518–10522

Chapter 5
Arrangement of Gold Nanoparticles onto a Slit-Type DNA Nanostructure in Various Patterns

5.1 Introduction

The programmed assembling of nanomaterials with functionality is one of the major goals for the bottom-up nanotechnology [1, 2]. Beyond the genetic carriers, DNA also have already been shown as one of the most promising polymers for the creation of various kinds of nanostructures, which are employed as scaffolds for the placing functional nanomaterials selectively and precisely into predesigned positions [1, 2]. The DNA origami technology, developed by Rothemund in 2006, provides a new strategy for the self-assembly of two-dimensional nanostructures [3–6], which have been utilized for the placement of various kinds of nanoparticles [7–16], programmed arrangement [17–21], molecular machines [22–24] and also been applied for the further design and construction of 3D architectures [25–30]. The placement of AuNPs onto various substrates always gains a lot of interests and the modified nanostructures are usually applied for the plasmonic waveguide and circuits. [31–33] For the further development of the AuNPs modified nanomartials in programmed patterns, new types of scaffolds are required and should be explored. Here a slit-type DNA origami carrying thiolated strands is designed and constructed, which is employed as the scaffold for the attachment of AuNPs at predesigned positions in various patterns.

5.2 Experimental Section

5.2.1 Preparation of Lipoic Acid Attached Oligonucleotides

N-Hydroxysuccinimide (NHS) ester of lipoic acid was synthesized according to a reported method. [34] NHS ester of lipoic acid dissolved in CH_3CN was added to amino-modified oligonucleotides in a solution containing 0.1 M Tris-HCl buffer

© Springer Japan 2015

Y. Yang, *Artificially Controllable Nanodevices Constructed by DNA Origami Technology*, Springer Theses, DOI 10.1007/978-4-431-55769-2_5

(pH 8.0) and CH_3CN (50 % volume), the reaction was carried out at room temperature for 4 h. Lipoamido-attached oligonucleotides were purified by a reverse-phase HPLC [JASCO LC2000 HPLC system, linear gradient, 2–30 % acetonitrile/50 mM ammonium formate, Nacalai Cosmosil C18 reversed-phase column (7.5 × 150 mm), 2.0 mL/min, 260 nm]. Lipoamido-attached oligonucleotides were collected, lyophilized, and redissolved in 10 mM Tri-HCl buffer (pH 7.6).

5.2.2 DNA Slit Preparation

4 μL 0.1 μM M13mp18 single stranded DNA (ordered from *New England Biolabs*), 10 μL 0.2 μM stapler strands (designed by caDNAno software), 4 μL 10-times buffer (200 mM pH 7.6 Tris-HCl, 10 mM EDTA and 100 mM $MgCl_2$) and 22 μL Milli-Q ultrapure water were all mixed together into tube. The mixture was then put in the PCR and annealed from 85 to 15 °C at a rate of-1.0 °C min^{-1}. The excess staple strands were removed by using gel filtration column (Sephacryl 300, GE Healthcare). The assembled structures were confirmed by AFM after the purification.

5.2.3 AuNPs Attachment onto the DNA Slit for Imaging

The 5 nm AuNPs were first pretreated with bis(p-sulfonatophenyl)-phenylphosphine (BSPP). The purified DNA slit was also treated with 1 mM TCEP at room temperature for 1 h to reduce the disulfide bond. The attachment was carried out by mixing the coated AuNPs (20 eq) and DNA slit solution at 4 °C for 4 h. In case of the attachment on the mica plate, the prepared sample (2 μL) was dropped onto a freshly cleaved mica plate for 5 min at room temperature and then washed three times using the same buffer solution. And then the sample was imaged by AFM.

5.3 Results and Discussion

5.3.1 Design and Preparation of DNA Slit

Using caDNAno software, a 2D DNA origami scaffold ($\sim$90 nm × $\sim$100 nm) having six parallel long and narrow spaces (each $\sim$5 nm × $\sim$60 nm) which look like the "slits" is shown in Fig. 5.1a, b. The cavities of the slits were designed for the placement of AuNPs ($\sim$5 nm) by using staple strands carrying thiol group orient the cavities. As a result, the six slit cavities can be employed as guide for arrangement of AuNPs in different patterns. As shown in Fig. 5.1a, the detailed design is that two thiolated staples with four thiol groups capture one AuNP at a specific position. And each staple strand carrying two thiol groups (Fig. 5.1c) will

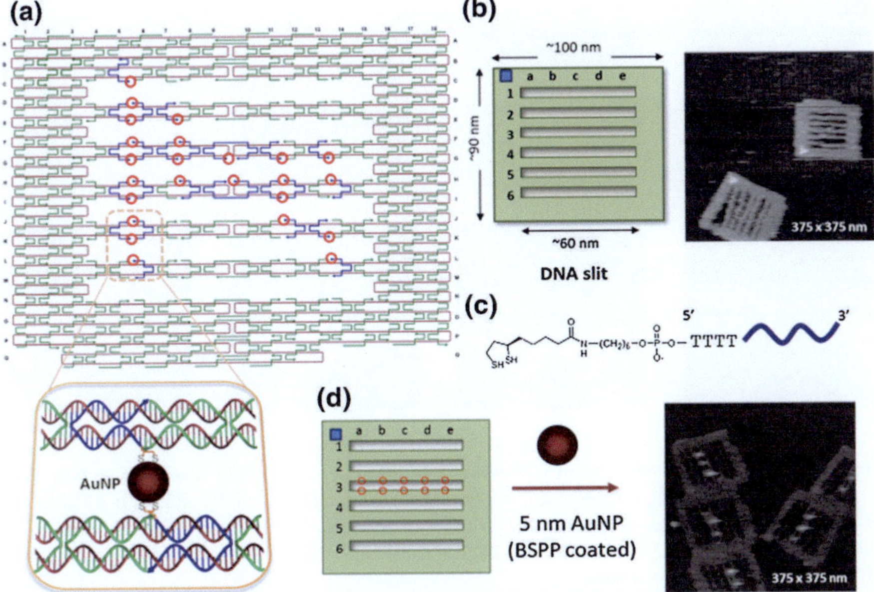

Fig. 5.1 Schematic design of DNA slit and introduction of AuNPs onto the silt scaffold in various patterns. **a** A rectangle DNA origami containing six long rectangular spcaces ($\sim$5 nm $\times$ $\sim$60 nm, in which the thiolated staple strands were circled by red color for the placement with AuNPs. **b** Simple drawing of DNA slit and the obtained AFM image. **c** Thiolated staple strand for the attachment with AuNPs. **d** AuNPs (5 nm, coated with BSPP) were introduced into thiolated DNA slit and imaged by AFM. [Endo, M.; Yang, Y.; Emura, T.; Hidaka, K.; Sugiyma, H. *Chem. Comm.*, **2011**, 47, 10743–10745.] Reproduced by permission of The Royal Society of Chemistry

be oriented to the cavity after assembling into the DNA slit, which will be convenient for the attachment with AuNPs in the predesigned positions. The AFM image shown in Fig. 5.1b confirmed that the DNA slit was successfully constructed. It can be seen that the six long and narrow cavities were clearly observed. The next step is to introduce of 5 nm AuNPs at predesigned positions. The selected staple strands along the cavities were modified by bis-thiol-group via an amino-linker and $(T)_4$-linker [8] (in Fig. 5.1d) and then incorporated together with other staples to form the DNA slit holding bis-thiol groups at the specific positions. The general preparation of AuNPs attached DNA slit is as follows: the pretreated 5 nm AuNPs were mixed with thiolated DNA slit and incubated at 4 °C for 4 h. The final nanostructures were all confirmed by AFM.

5.3.2   Placement of AuNPs onto DNA Slit in Solution

The DNA slit in the horizontal arrangement of AuNPs was first examined, in which five AuNPs were expected to attach only along one of the slits (Figs. 5.1d and 5.2).

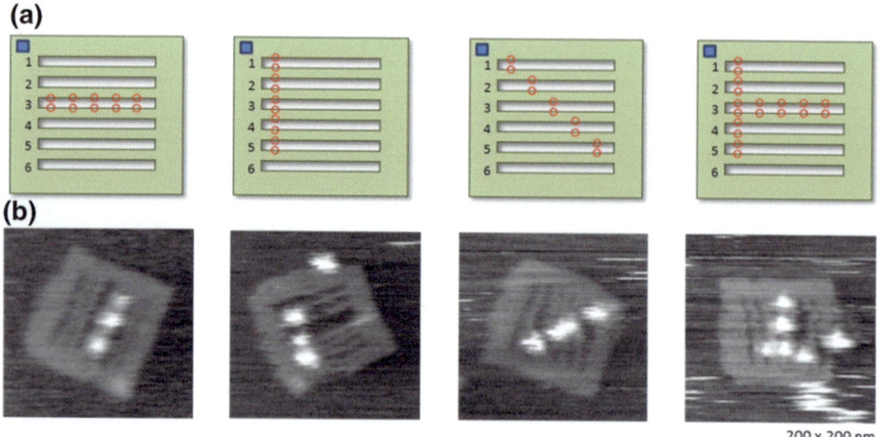

Fig. 5.2 Introduction of AuNPs onto DNA slit in solution into four kinds of patterns: **a** Four kinds of arrangements of AuNPs on DNA slit: horizontal, vertical, diagonal and T-shape. **b** AFM images of the DNA slit after the treatment with AuNPs in solution. [Endo, M.; Yang, Y.; Emura, T.; Hidaka, K.; Sugiyma, H. *Chem. Comm.*, **2011**, 47, 10743–10745.] Reproduced by permission of The Royal Society of Chemistry

From the AFM image in Fig. 5.2, it can be seen that AuNPs were located successfully in the slit cavity, but most of the DNA slits were attached with three AuNPs ($\sim$80 % yield) which were distributed equally along the cavity but far from the both left and right sides. The incomplete placement of pre-set five AuNPs is probably because the $(T)_4$ linker of staples is long enough to reach the middle of adjacent positions resulting that more than two bis-thiol group capture one AuNPs even though the distance (10 nm) of the adjacent bis-thiol-staple is already enough for the placement of five 5 nm AuNPs along one cavity. Besides using the DNA slit containing thiol group at both sides of the slit for the attachment of AuNPs, one-side bis-thiol DNA slit was also investigated. As a result, the yield of three-AuNP attachment was decreased compared to two-side bis-thiol DNA slit and some AuNPs attached on the boundary of one side of the slit but in the slit cavity, indicating that the two-side bis-thiol groups were necessary to guarantee the attachment of AuNPs at the desired positions. These results show that the two-side thiolated DNA slit design works effectively for guiding the correct placement of AuNPs. Furthermore, other patterns of AuNP attached on DNA slit were also carried out at the same time. The results were all shown in Fig. 5.2. Not only the horizontal pattern, but also vertical, diagonal and T-shaped patterns were obtained.

5.3.3   *Placement of AuNPs onto DNA Slit on Mica Surface*

Besides the assembling in solution, it was also tried to trap the AuNPs onto DNA slit directly on the mica surface, where the DNA slits were all fixed because of the electrostatic interaction. After the adsorption of thiolated DNA slit on the mica surface, a drop of AuNPs solution was added to the pretreated mica plate and incubated for a few minutes. After washing the mica plate with buffer and remove the excess AuNPs, the sample was examined by AFM. Here same four patterns of arrangement of AuNPs were prepared (Fig. 5.3): horizontal, vertical, diagonal and T-shape. From the AFM images, the five-AuNP attachment was obtained in horizontal, vertical and diagonal patterns, indicating that two bis-thiol groups of two adjacent staple strands were close enough to trap one AuNPs in the cavity. However, the yield of five-AuNP in horizontal pattern is low and main products are four-AuNP and three-AuNP. The yields of each pattern containing different numbers of AuNPs were summarized together in Table 5.1. A T-shape pattern was tried to assemble by mixing both of horizontal and vertical arrangement. As a result,

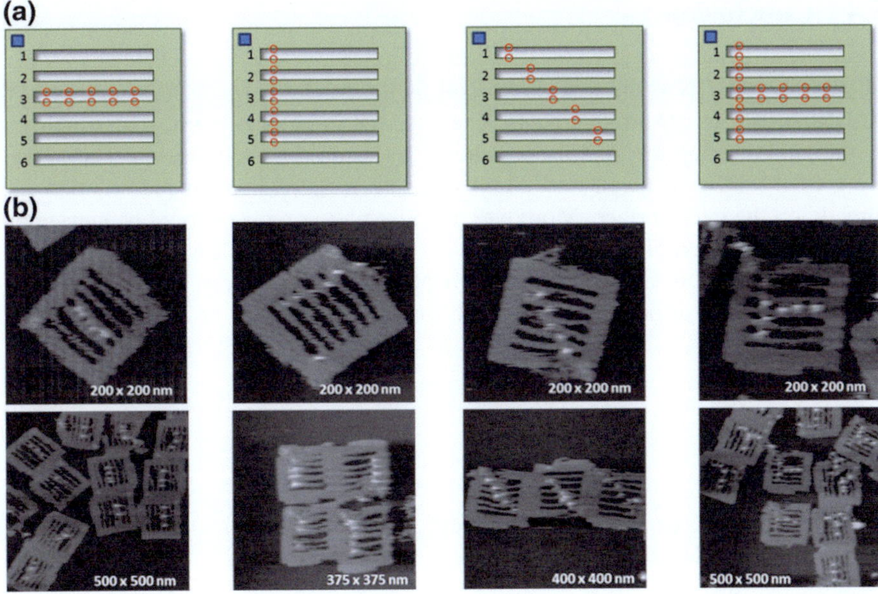

Fig. 5.3 Direct placement of AuNPs onto DNA slit on mica surface. a. Four kinds of arrangements of AuNPs along different cavities in DNA slit: horizontal, vertical, diagonal and T-shape. Positions for thiolated staple strands were circled by red color for the placement of AuNPs. b. AFM images of AuNP-attached DNA slits into four kinds of predesigned patterns. Images correspond to the upper schematic drawings of DNA slits. [Endo, M.; Yang, Y.; Emura, T.; Hidaka, K.; Sugiyama, H. *Chem. Comm.*, **2011**, 47, 10743–10745.] Reproduced by permission of The Royal Society of Chemistry

Table 5.1 Ratio of attachment of AuNPs to the DNA slits containing thiolated staples

	Number of AuNPs				
	5	4	3	<2	Number of slits counted
Horizontal	7 %	41 %	46 %	6 %	61
Vertical	61 %	27 %	11 %	0 %	62
Diagonal	62 %	31 %	7 %	0 %	74
	Number of AuNPs				
	9	8	7	6	<5	Number of slits counted
T-shape	0 %	26 %	42 %	27 %	5 %	106

[Endo, M.; Yang, Y.; Emura, T.; Hidaka, K.; Sugiyama, H. *Chem. Comm.*, **2011**, 47, 10743-10745.] Reproduced by permission of The Royal Society of Chemistry

the horizontally and vertically assembled AuNPs was up to eight-AuNPs. Unfortunately, fully assembled T-shape (total nine-AuNPs) was not obtained because of the low yield of five-AuNP in horizontal pattern.

5.4   Conclusion

In summary, a DNA slit was developed as the scaffold for the assembling with AuNPs in various 2D arrangements. The silt cavities in the origami were employed as the guide to trap the AuNPs in patterns by using thiolated staples. The assembling of the DNA origami and AuNPs can be realized both in the solution and directly on the mica surface. Moreover, the assembling process could be carried out easily and the assembled nanostructures were all confirmed clearly by using AFM. A simple programmed assembly system of nanomaterials was successfully illustrated here, which afford a new approach for the investigation of new AuNPs based nanomaterials.

References

1. Niemeyer CM, Mirkin CA (2004) Nanobiotechnology: concepts, applications and perspectives. Wiley-VCH, Germany
2. Endo M, Sugiyama H (2009) ChemBioChem 10:2420–2443
3. Rothemund PWK (2006) Nature 440:297–302
4. Kuzuya A, Komiyama M (2010) Nanoscale 2:310–322
5. Torring T, Voigt NV, Nangreave J, Yan H, Gothelf KV (2011) Chem Soc Rev 40:5636–5646
6. Rajendran A, Endo M, Sugiyama H (2012) Angew Chem Int Ed 51:874–890
7. Ke Y, Lindsay S, Chang Y, Liu Y, Yan H (2008) Science 319:180–183
8. Shen W, Zhong H, Neff D, Norton ML (2009) J Am Chem Soc 131:6660–6661
9. Stephanopoulos N, Liu M, Tong GJ, Li Z, Liu Y, Yan H, Francis MB (2010) Nano Lett 10:2714–2720
10. Endo M, Katsuda Y, Hidaka K, Sugiyama H (2010) J Am Chem Soc 132:1592–1597

11. Sannohe Y, Endo M, Katsuda Y, Hidaka K, Sugiyama H (2010) J Am Chem Soc 132:16311–16313
12. Endo M, Katsuda Y, Hidaka K, Sugiyama H (2010) Angew Chem Int Ed 49:9412–9416
13. Sharma J, Chhabra R, Andersen CS, Gothelf KV, Yan H, Liu Y (2008) J Am Chem Soc 130:7820–7821
14. Ding B, Deng Z, Yan H, Cabrini S, Zuckermann RN, Bokor J (2010) J Am Chem Soc 132:3248–3249
15. Hung AM, Micheel CM, Bozano LD, Osterbur LW, Wallraff GM, Cha JN (2010) Nat Nanotechnol 5:121–126
16. Kuzuya A, Koshi N, Kimura M, Numajiri K, Yamazaki T, Ohnishi T, Okada F, Komiyama M (2010) Small 6:2664–2667
17. Endo M, Sugita T, Katsuda Y, Hidaka K, Sugiyama H (2010) Chem Eur J 16:5362–5368
18. Zhao Z, Yan H, Liu Y (2010) Angew Chem Int Ed 49:1414–1417
19. Li Z, Liu M, Wang L, Nangreave J, Yan H, Liu Y (2010) J Am Chem Soc 132:13545–13552
20. Rajendran A, Endo M, Katsuda Y, Hidaka K, Sugiyama H (2011) ACS Nano 5:665–671
21. Endo M, Sugita T, Rajendran A, Katsuda Y, Emura T, Hidaka K, Sugiyama H (2011) Chem Commun 47:3213–3215
22. Gu H, Chao J, Xiao S-J, Seeman NC (2010) Nature 465:202–205
23. Lund K, Manzo AJ, Dabby N, Michelotti N, Johnson-Buck A, Nangreave J, Taylor S, Pei R, Stojanovic MN, Walter NG, Winfree E, Yan H (2010) Nature 465:206–210
24. Wickham SFJ, Endo M, Katsuda Y, Hidaka K, Bath J, Sugiyama H, Turberfield AJ (2011) Nat Nanotechnol 6:166–169
25. Douglas SM, Chou JJ (2007) W. M. Shih. Proc Natl Acad Sci USA 104:6644–6648
26. Andersen ES, Dong M, Nielsen MM, Jahn K, Subramani R, Mamdouh W, Golas MM, Sander B, Stark H, Oliveira CLP, Pedersen JS, Birkedal V, Besenbacher F, Gothelf KV, Kjems J (2009) Nature 459:73–76
27. Douglas SM, Dietz H, Liedl T, Hogberg B, Graf F, Shih WM (2009) Nature 459:414–418
28. Dietz H, Douglas SM, Shih WM (2009) Science 325:725; Kuzuya A, Komiyama M (2009) Chem Commun 2009:4182–4184
29. Ke Y, Sharma J, Liu M, Jahn K, Liu Y, Yan H (2009) Nano Lett 9:2445–2447
30. Endo M, Hidaka K, Kato T, Namba K, Sugiyama H (2009) J Am Chem Soc 131:15570–15571
31. Ozbay E (2006) Science 311:189–193
32. Jones MR, Osberg KD, Macfarlane RJ, Langille MR, Mirkin CA (2011) Chem Rev 111:3736–3827
33. Tan SJ, Campolongo MJ, Luo D, Cheng W (2011) Nat Nanotechnol 6:268–276
34. Sharma J, Chhabra R, Andersen CS, Gothelf KV, Yan H, Liu Y (2008) J Am Chem Soc 130:7820–7821

Curriculum Vitae

Dr. Yangyang Yang
Institute for Integrated Cell-Material Sciences,
Kyoto University, Yoshida Ushinomiya-cho,
Sakyo-ku, Kyoto, Japan, 606-8501
Tel.: +81-75-753-9770
Fax: +81-75-753-9785
e-mail: triyang@kuchem.kyoto-u.ac.jp

Education

Kyoto University
Degree: Ph.D. March 2014
Thesis advisor: Prof. Hiroshi Sugiyama; Assoc. Prof. Masayuki Endo
Thesis topic: Artificially controllable nanodevices constructed by DNA origami
technology: photofunctionalization and single molecule analysis

East China University of Science and Technology
Degree: M.S. March 2011
Thesis advisor: Prof. Yufang Xu; Prof. Xuhong Qian
Thesis topic: New Fluorescent Probes based on Biological applications: Design,
Synthesis and Application

Yangzhou University
Degree: June 2008, B.S.
Major: Chemistry.

© Springer Japan 2015

Y. Yang, *Artificially Controllable Nanodevices Constructed by DNA
Origami Technology*, Springer Theses, DOI 10.1007/978-4-431-55769-2

Major Honors and Awards

- JSPS Research Fellowship for Young Scientists: 2013–2015
- TakekoshiSho of Kyoto University: Kyoto University, 2013
- Outstanding Graduate Student: Yangzhou University, 2008
- Undergraduate Student Scholarship: Yangzhou University, 2005

Research Interests

I am very interested in chemical biology, in which especially the dynamic behaviors of biomolecules gained much of my attentions. As a doctorate candidate, I mainly worked on DNA-based nanomaterials. It is so fascinating that the genetic carriers, DNA molecule, can be applied for the construction of nanometer-sized materials. Taking advantages of single molecule analysis methods such as high-speed AFM, it has been successfully observed various kinds of molecular motions of biomolecules based on DNA nanotechnology. These DNA-based biomaterials show great potentials for the further applications in various fields like nanophotonics, nano-medicine and so on. Right now I am focusing on the development of 3D DNA nanostructure for target molecule delivery on biosystems.